DISTINGUISHED MEN AND WOMEN OF SCIENCE, MEDICINE AND THE ARTS

HAMILTON'S HISTORY OF MEDICINE AND SURGERY

DISTINGUISHED MEN AND WOMEN OF SCIENCE, MEDICINE AND THE ARTS

Additional books in this series can be found on Nova's website under the Series tab.

Additional e-books in this series can be found on Nova's website under the e-book tab.

Distinguished Men and Women of Science, Medicine and the Arts

Hamilton's History of Medicine and Surgery

R. Shane Tubbs
Anand N. Bosmia
Meghan Richli
Marios Loukas
Paul S. Agutter
and
Mohammadali M. Shoja

Nova Biomedical
New York

Library of Congress Cataloging-in-Publication Data

Hamilton, William, Sir, 1788-1856, author.
Hamilton's History of medicine and surgery / [edited by] R. Shane Tubbs, Marios Loukas, Mohammadali M. Shoja and Paul Agutter (Pediatric Neurosurgery, Children's Hospital, Birmingham, AL, USA).
pages cm
Includes index.
ISBN: 978-1-62948-882-0 (hardcover)
1. Medicine--History. 2. Surgery--History. 3. Anatomy--History. I. Tubbs, R. Shane, editor. II. Title. III. Title: History of medicine and surgery.
R131.H222 2014
610--dc23

2013048016

Published by Nova Science Publishers, Inc. † New York

To my parents, Richard and Karon Tubbs

R. Shane Tubbs, PhD

To Natwarlal Bosmia, my beloved father whom I sorely miss, and to Beena Bosmia, my beloved mother whose fortitude and grace have sustained me

Anand N. Bosmia, BA

To Dr. R. Shane Tubbs, for allowing me the opportunity to be involved in this project and for his unwavering patience; to Dr. Marios Loukas, for the kindness and faith he has bestowed upon me in recommending me for this position; to Dr. Mohammadali M. Shoja, for his helpful advice, generosity, and friendship during my time at Children's of Alabama; and lastly to my parents, for their endless love and support

Meghan Richli, BS

To my parents, Anastasia and Christos Loukas

Marios Loukas, MD, PhD

To Alun H. Maddy, who taught me to think scientifically and to question what is commonly assumed, and to Lewis Thomas, who taught me how to balance caring with common sense

Paul S. Agutter, MD, PhD

To Susan Tubbs, for all her caring and moral support, and to Alireza Ghaffari, for his wonderful mentorship

Mohammadali M. Shoja, MD

Contents

Editors' Preface

Knowledge of the history of medicine enriches our understanding and appreciation of the present as well as the past, and enhances both research and clinical practice. Reading Hamilton's *History of Medicine, Surgery and Anatomy*, hitherto a rather obscure classic, can alter the perspectives of twenty-first century readers in three distinct but complementary ways.

First, the book is a minor masterpiece of early nineteenth century scholarship and erudition. Hamilton's facility with classical languages was matched by his broad and detailed grasp of the major contributions to medical knowledge, theory and practice from the dawn of history to his own time. For anyone who seeks outline information about the work of a physician or surgeon from the more or less remote past, Hamilton's book is likely to provide it. Furthermore, this text is likely to place that information in a historical context more quickly and efficiently than an online search would.

Second, the work is itself a historical document, and, in certain respects, a primary source. It was written when physiology and pathology were in their infancy. The world was yet to be graced with the discoveries of Cruveilhier, Andral, Virchow, du Bois Reymond, Ludwig, Bernard, and their eminent contemporaries and successors. Lister's achromatic lens had improved the microscope's design, but in the absence of staining techniques, knowledge of tissue morphology remained primitive. Cell theory was not yet born; Raspail's studies had recently adumbrated it, but they were not known outside of France, and almost a decade was to elapse before Schwann and Schleiden popularized the concept of the "living cell". Wöhler and Liebig had only recently dug the foundations of organic chemistry, and biochemistry would not become a word, much less a discipline, for another seventy years. No one knew the causes of infection, or yet had heard names such as Pasteur and Koch; "inflammation" was a vague idea, and physicians routinely treated fevered patients with bloodletting. It is salutary to read a book that portrays the history of medicine from a perspective so very different from that of the present century; it can startle, and sometimes amuse, its readers, but above all, such a text increases our appreciation of the advances we have made during the past two hundred years.

The third point is less obvious. True to the culture of his time, Hamilton took an approach to history that modern historians of science and medicine would eschew. In reflecting on what we see as the shortcomings of this approach, we improve our capacity for balanced criticism. For one thing, Hamilton's viewpoint was that of "Whig history". He saw the history of medicine as a more or less steady progression from the darkness of ignorance and superstition to the light of reason and understanding. That view of the evolution of knowledge is simplistic, to say the least, but it would be absurd to blame Hamilton for it; everyone at the time (and for the following century or more) took the same approach. By the same token, he holds an unshakeable belief in the rightness of Christianity and the superiority of European (specifically British) culture, which, like his contemporaries, he tacitly assumes to be the pinnacle of civilization. For another, Hamilton's book is a "great men" portrayal of the past, every advance in theory and practice is attributed to the genius of one or more eminent individuals. Although such a portrayal cannot be deemed "wrong", it is simplistic and entails a limited understanding of history. Advances in medicine do not take place in isolation. They are facets of general cultural change, and advances in one area inform another. Hamilton's book repeatedly illustrates this limitation of the "great men" portrayal. To choose one example, when in chapter 11 he describes the seventeenth century work of Thomas Sydenham and the pioneers of the Royal Society, he praises them for eschewing speculation and hypothesis and relying instead on observation and induction. However, he never mentions that this approach to natural philosophy resulted entirely from Francis Bacon's teaching published earlier in the century. Thus, Sydenham and his contemporaries were applying Bacon's principles of knowledge-acquisition to their own fields of endeavor to the best of their ability. Once we recognize that, we are in a better position to read Hamilton's book with the critical appreciation it merits and thereby enhance our critical faculties to the lasting benefit of our own studies and practice.

We are indebted to Dr. E.C. Nelson for information about the author of the original text.[1] According to the limited biographical data available, William Hamilton was probably born in the remote parish of Clondavaddog, County Donegal, Ireland, around 1783 (some evidence suggests that the year of his birth could have been as late as 1786). A letter he wrote to William Hooker in 1824 indicates that he was one of the thirteen children of the Rev. Dr. William Hamilton of Clondavaddog, who had written the first geological account of the Giant's Causeway on the north coast of Ireland and was the victim of a political murder in 1797. William Junior enrolled at Trinity College, Dublin, in 1799 and received a Bachelor of Arts degree in 1805. He then entered Magdalene College, Oxford, and graduated in 1809 with a Master of Arts degree and a Bachelor of Medicine degree. He settled in Plymouth, Devon, England, and wrote the *History of Medicine, Surgery and Anatomy* while residing there, as indicated at the end of his preface, and his name is recorded as a Plymouth resident in the census surveys of 1841 and 1851.[2] Most of his work was in the field of botany, and he seems to have been an avid collector of plants, especially of those with medical value. This interest led him to travel at least once (probably twice) to the West Indies and likely also to Mexico. Apart from the *History of Medicine, Surgery and Anatomy* he is only known to have written one other medical work, which is rare and obscure.[3]

[1] See, Nelson, E. Charles (1994), "Dr William Hamilton (ca. 1783-1856), botanist, of Donegal and Plymouth." *Archives of Natural History* 21(1), 73-76, and references cited therein, notably the bibliographical compilation by Howard *et al.* Dr Nelson's correspondence with us during our work on Hamilton's book was greatly appreciated. We are also indebted to Alison Nelson of the Bodleian Library, Oxford, for putting us in touch with him and for helping us to trace original sources.

[2] British census surveys are conducted once every 10 years; 1841 was the first.

[3] Hamilton, W. (1810), *The Encheiridion medicum; or, Young Practitioner's Pocket Companion, being a conspectus of the London, Dublin, and Edinburgh Pharmacopoeias*. London, Longman, Hurst, Rees & Orme. Cited by Nelson (*op. cit.*)

In preparing this new edition, we have tried to balance respect for the original text with the needs and interests of the modern reader. We have abridged the text, deleting most of Hamilton's rhetorical flights and repetitions, which are characteristic of English non-fiction written during the early nineteenth century, but seeking nevertheless to retain the style and character of the original text. Some of the more aggressively ethnocentric passages have been removed, since they add nothing of value to the historical information in the book. The British English of the period has been replaced with modern American English, but we have retained much of the original punctuation. Some of Hamilton's copious footnotes have been retained, others incorporated into the text, and others removed altogether. We have added our own footnotes as well. Hamilton's occasional spelling errors in Latin have been corrected. In the end, the current work hopes to make an obscure and more or less forgotten masterpiece readily accessible and comprehensible to the modern reader.

For those who wish to read the original book, at least for comparison, a two-volume version can be found online at http,//archive.org.

Sir William Hamilton's Original Preface

However arid and uninviting the prospect of a history of medicine may appear at a distance, it will be found gradually to improve, and become full of interest, wonder, and animation, as we proceed. The history of medicine is, in fact, the history of the human species, uncontaminated by those crimes and disorders, which blot the page of other histories. The history of medicine, on the contrary, is the history of peace and good will, of endless harmony, and unceasing philanthropy. Instead of recording the ruins of war, and the growth of immorality, her province is to note the lessening of man's suffering; and the only triumphs which she records are those obtained over sickness, death, and sorrow.

Lost in the depths of unrecorded ancient time, the earlier pages of medical, as well as general history are few, obscure, and unsatisfactory. Like travelers enveloped in an Alpine mist, we wander in darkness and uncertainty, without compass to direct, or planet to enlighten us. The same darkness, doubt, and uncertainty, mark our progress through the earlier periods of Egyptian and Grecian history, and we find ourselves bewildered in a motley crowd of names who bear so little similitude to substantial personages, that we seem to be traversing the land of dreams and the regions of romance, rather than those of real history and actual events. It is not until we approach the confines of Christianity that the darkness begins to lessen in intensity. Hippocrates is the first who stands in tolerable distinctness before us, actively striving to eliminate the mists of ignorance, and rescue the art of healing from the hands of fraudulence and superstition.

With the birth of Christianity, the human mind began to feel and exert its powers. But in the very infancy of this new-born day, an immense darkness once more involved the world, and the constraints of ignorance now connected to the earth the celestial mind of man, until the discovery of the art of printing newly unbarred the gates of Heaven, and let in that flood of knowledge, which enabled men to emancipate themselves again from the slavery of superstition and prove themselves to be inferior only to the Angels. Such is the nature of a history of medicine; and if the journey be commenced in darkness and perplexity, it terminates at least in the regions of splendid brilliancy and joyous certainty.

In performing this long but interesting journey, the incidents, which were few and scattered at the commencement, thicken as we advance; the landscape, which was at first but dimly seen, becomes distinct and bright, and the glory of the morning reveals new discoveries for our surprise. If, amidst the multitude of objects which claim our notice, some have been overlooked, the goodwill of the reader will surely not go unnoticed.

Sir William Hamilton, 1831

About the Editors

R. Shane Tubbs, PhD, is an anatomist at Children's of Alabama, Birmingham, Alabama. He is a Professor of Anatomy in the Department of Anatomical Sciences, St. George's University, Grenada, and in the Centre for Anatomy and Human Identification, University of Dundee, Scotland, United Kingdom. His research focuses on anatomy and translational research, in particular the application of anatomy to neurosurgical problems.

Anand N. Bosmia, BA, is a medical student at the UAB School of Medicine, Birmingham, Alabama. He is an associate editor of the International Journal of History and Philosophy of Medicine. His main interests are the history of medicine and surgery, the intersection of religion and medicine, and cults and new religious movements.

Meghan Richli, BS, is a medical student at St. George's University, Grenada, who contributed to this book during her research fellowship at Children's of Alabama. She is currently completing her clinical rotations in Brooklyn, New York, and has a continued interest in taking part in medical research.

Marios Loukas, MD, PhD, is the Chair of and Professor of Anatomy in the Department of Anatomical Sciences, St. George's University, Grenada. He is also the Dean of Research for the School of Medicine at St. George's University. His research focuses on anatomy and translational research, medical education and assessment, and ultrasound.

Paul S. Agutter, MD, PhD, is the founder of Theoretical Medicine and Biology Group, Glossop, Derbyshire, United Kingdom. He was formerly editor-in-chief of Theoretical Biology and Medical Modelling. He has authored *Thinking about Life, The History and Philosophy of Biology and Other Sciences*, *About Life, Concepts in Modern Biology*, and several other books.

Mohammadali M. Shoja, MD, is a research fellow in the Section of Pediatric Neurosurgery, Children's of Alabama, and in the Department of Neurosurgery, University of Alabama at Birmingham. He is one of the founding members of the Medical Philosophy and History Research Center at Tabriz University of Medical Sciences, Tabriz, Iran. He is editor-in-chief of the International Journal of History and Philosophy of Medicine. He has authored and edited multiple textbooks, including *Hypotheses in Clinical Medicine* and *A Practical*

Guide to the Dissection of Head and Neck with Descriptive Anatomy. His primary research interests include the history and philosophy of medicine, neurosurgical anatomy, brain tumors, and cerebrovascular diseases, in particular aneurysm and vasospasm.

Chapter 1

History of Medicine from the Creation to the Birth of Hippocrates

Tracing the Dawn of Medical Science

It might have been expected that the origin of a branch of knowledge as conducive and essential to the welfare and the preservation of the human race as medicine would have been traceable to its remotest source, with a precision equal to its importance. The reverse, however, is the case, and the farther we attempt to penetrate into the mists of antiquity, the more indistinct does the object of our pursuit become. We are justified, however, by many circumstances recorded in the Holy Scriptures in believing that medicine and many other useful branches of knowledge had arrived at a very considerable degree of perfection before the Flood. Whether man acquired this knowledge through an immediate revelation from the Almighty or cultivated it through the exercise of his own judgment and accidental discoveries is a question that human reason cannot hope to answer.

Conjectures of Schulze and Others Concerning Medicine in Ancient Times

Schulze, who served as a professor at Altdorf in the early eighteenth century, traces the origin of medicine to the period of the Fall. He points out the strong probability that Adam, yielding to the all-authoritative voice of necessity, discharged the office of an obstetrician to his wife, and thus performed the first surgical operation.[1] The *History of Medicine* by the French writer Le Clerc provides a copious view of the doctrines of the ancients down to the days of Galen Le Clerc discusses with considerable ingenuity the question of whether the science of medicine emanated directly from heaven. He also traces the practice of medicine in

[1] J.H. Schulzii Historia Medicinæ a rerum initio ad A.U.C. 535, deducta. 4to. Lipsiæ, 1728. Ejusdem Compendium Historiæ Medicæ, a rerum initio ad Hadriani accessum. 8vo. Halæ, 1742.

its various branches to the days of Adam, contending that Adam was the first physician, surgeon and obstetrician.[2]

Brambilla, a renowned and talented surgeon who practiced for many years in Vienna and was appointed by Emperor Francis the Second as the head of the Academy of Surgery founded in Vienna in 1783, traces the invention of surgical instruments to Tubal Cain, the great-grandson of Adam. Per Genesis 4,22, Tubal Cain was "an instructor of every artificer in brass and iron", and hence not unlikely to have been the first to imagine and make such simple instruments as surgical knowledge in those early ages required.

In referring the origin of surgical instruments to a period as remote as the second generation from Adam, it must not be supposed that men were very proficient at the art of surgery in those early days. As medical knowledge prior to the Flood may be reasonably presumed to have been confined to the use of a few plants of the mildest character and most innocuous activity, so we may conclude that surgical skill was restricted to the treatment of the most ordinary wounds and the application of the least complicated bandages, to the reduction of a dislocated joint and the setting of a broken limb. The amputation of a gangrenous leg or the trepanning of a fractured skull would have required skill that transcended medical knowledge prior to the Flood.

Circumcision, The First Surgical Operation on Record

It was not until long after the Flood that the Almighty found it necessary to select another Noah from among the lineage of Adam, in the person of Abraham, and to distinguish him and his descendants as a people set apart for himself by the ordinance of a peculiar operation performed upon the males, circumcision. This occurrence furnishes the earliest recorded instance of a surgical operation. With the solitary exception of the surgical rite of circumcision, history does not provide us with any information concerning the progress of either medicine or surgery during the patriarchal ages. We can, however, readily believe that the practice of both must have been of the simplest description, for disease had not yet become so complex in character that it demanded the utmost skill in diagnostics or acquired the virulence that too often baffles the utmost efforts of art. In ancient times, man's food was simpler in its character, and his habits more in accord with the dictates of nature.

Medicine among the Ancient Egyptians

We find the study of medicine first cultivated with method and reduced to something approaching systematic arrangement among the inhabitants of Egypt. The Egyptians were undoubtedly further ahead than other primitive nation in the march of civilization and the cultivation of knowledge. Anatomy has been imagined by some to have been practiced by the Egyptians at a period long before the earliest existing records. Manetho, a distinguished

---

[2] Histoire de la Médicine, où l'on voit l'Origine et les progrès de cet Art, avec figures, par Daniel le Clerc. 12mo. Geneva, 1696. 4to. Part 3me. Amsterdam. 1725.

Egyptian writer, is represented by Eusebius as stating that Athotis, an Egyptian monarch, wrote several treatises on anatomy. However, when we learn that the era assigned to Athotis would carry us back to an age prior to the formation of Adam, we can easily estimate the degree of credibility to which such a fable is entitled.

It is not improbable that accidental circumstances may have given men some imperfect notion of the internal structure of the human frame, or that inspecting the viscera of animals that they killed may have led them to infer analogies concerning the disposition and functions of the corresponding parts in the human body. However, few would agree that such knowledge approached accurate anatomical science. Human dissection very likely did not commence before the days of Erasistratus and Herophilus, but a few centuries before the birth of Christ, or, at the earliest, before the days of Hippocrates. Prior to this period, human beings were superstitious towards the act of human dissection, which was viewed as offensive toward the gods. Thus, men were unable to acquire a sophisticated knowledge of the organization and physiology of the human viscera and thereby, could not determine the causes of disease or adopt a rational method of cure. They were disposed to regard every complaint whose origin they could not explain or whose obstinacy they could not address with their skill as the result of demonic influence or celestial wrath.

We can readily understand how almost every nation of antiquity came to refer the origin of medicine to the immediate instruction of the supreme powers. According to the traditions of Egypt, the first collection of medical and other precepts was derived from Thouth or Taaut, who, according to Diodorus, was the secretary of Osiris and a man of the most varied talents and almost unearthly skill. This extraordinary icon flourished nearly two thousand years before the commencement of the Christian era. The majority of the precepts received from Thouth related to the practice of medicine, and these precepts were collected into a volume entitled *Embre*. This constituted the textbook of the priests, who were the only physicians in those early days and were required to follow it in their practice as a sacred and unfailing guide. The ancient Egyptians thought it better that ninety-nine patients should fall victims to an erroneous practice than the validity of the instructions delivered by the mighty Thouth be questioned or one heretical patient presume to recover in contradiction to them. Next after Thouth in the Egyptian annals of medicine, we find Apis and Serapis recorded among the earliest founders or improvers of medicine and, like Thouth, worshipped after their deaths. Apis had been a king of the Argives, and his zeal for the improvement of his fellow men led him to Egypt for the purpose of instructing its inhabitants in the arts of civilized life. Apis so ingratiated himself with his pupils that they raised him to the throne of their country and worshipped him after his death. Apis flourished about 1,745 years before Christ, and was regarded by some as the inventor of medicine. His reputation, however, fell far short of that of Serapis, to whom divine honors were paid by both the Egyptians and the Greeks.

Since the art of healing was viewed as celestial in origin, it was practiced exclusively by the priests, whose presumed influence with the inhabitants of heaven naturally marked them as the most potent defense in the combat with disease. To secure the permanence of this monopoly, the priests strove to disguise the rules of their practice beneath a multitude of superstitious observances and surround it with a fence of imposing and impenetrable mystery. Magic and divination were regarded as the highest branches of the profession and were especially reserved for the chief priests, whereas the study of the hermetical books on medicine was left for the Pastophori, or image bearers, who held a very subordinate rank within the priesthood. The Pastophori were assigned the task of applying the remedies prescribed in the sacred volumes of Thouth. Their superiors in rank functioned to exercise superhuman powers and thereby direct the course of the malady, and to predict the issue by their prophetic skill.

It should come as no surprise that medical knowledge remained stationary for so long. The office of these priests was hereditary and their privileges were exclusive. Thus, the exclusion of others closed the door for competition among practitioners of medicine and offered no opportunity for the display of relatively superior skills. Improvement of accepted practices was arrested by punishment for attempting to deviate from the fundamental principles laid down in the volumes ascribed to Thouth. Furthermore, the interests of the priests perpetuated the popular delusions concerning the cause of disease and the necessity of attaining the favor of the gods through prayer and appeasing their ministers. To note, we can readily believe that the extent of the priests' skill consisted in accurately watching, and carefully seconding, the indications furnished by nature to direct treatment, a conjecture strengthened by Aristotle's recording of the fact that it was contrary to their practice to interfere with the course of the patient's illness earlier than the fourth day after its onset. Thus, if the patient did not die or recover, the priests secured sufficient opportunity to watch the progress of the illness and identify the outcome. This practice resulted in further confirmation of their own skill and pretended influence with the gods.

Around the time of Herodotus, about four and a half centuries before the Christian era, a revolution in the practice of medicine among the Egyptians appears to have taken place. At this time, every distinct malady was followed by a physician who limited his efforts to the study and treatment of that disease alone. This description, given by Herodotus, is hardly applicable to the "priest physicians" discussed previously and suggests that various esoteric practitioners began about this time to challenge the priests' popularity.

From the manner in which Herodotus speaks of this distribution of several divisions of medical practice, it might be imagined that the practitioners of each would have become adept at diagnostics and treatment within their respective fields. However, when we recollect the holy horror with which the anatomy of the human body continued to be regarded and the various other difficulties that these specialists had to surmount in their pursuit of professional knowledge, we can believe without much difficulty that they were little, if at all, superior to their spiritual rivals with regards to practical skill. However deficient they were in surgical skill, their acquaintance with vegetable simples appears to have been considerable and their pharmaceutical knowledge to have extended to the more active properties of some of the most powerful articles of the modern materia medica.

Citing Genesis 50,2, in which Joseph orders "his servants the physicians to embalm his father [Jacob]", some writers erroneously concluded that the practice of medicine was followed as a distinct profession in Egypt as early as before the death of Jacob, or about 1700 years before the Christian era. However, when we consider the manner in which the process of embalming was conducted, the small estimation in which operators of this description were held, the great subdivision of labor for which the Egyptians were remarkable, and the severe restrictions imposed upon the interference of persons of one profession or calling with those of another, we cannot but admit the improbability that any physician would degrade himself to perform what was regarded as a menial operation within ancient Egyptian society. We have no sufficient warrant for concluding from this passage that medicine began to be practiced as a distinct profession in Egypt at any time prior to the period assigned by Herodotus, which must have been at least 1,200 years later than the death and embalming of Jacob.[3]

Medicine among the Ancient Jews

The Jews, upon leaving the Egyptian territories, in which they had resided for 430 years, continued to observe the customs and institutions of the Egyptians, including the practice of medicine. It is no surprise that the priests among the Jews attributed to themselves the whole of this practice, as the priests of the Egyptians did the same. The religious history of the Jews is laden with instances in which the interposition of God led to the infliction and removal of diseases. Various persons in Biblical scripture were smitten with disease for disobedience towards God or one of his emissaries, such as Miriam with leprosy for speaking against the authority of Moses.

Medicine in Ancient Hindustan

The proficiency of the inhabitants of the Indian subcontinent, or Hindustan, in the art of healing and medicine does not appear to have equaled that of nations far less enlightened in other disciplines. According to Strabo, they chiefly relied on a rigid attention to diet and the external application of cataplasms and other topical remedies to cure disease. Medicine appears to have been practiced chiefly, if not entirely, by persons called Samaneans, who exercised their calling by special permission from, and under the immediate supervision of, the magistrates. However, their practice of medicine was encumbered with many difficulties, and the spirit of useful inquiry necessary for medicine to progress was repressed by various laws. Practitioners of medicine were expected to know and provide an antidote to any medication that they prescribed in the event that that particular medication made a patient ill. Failure to do so was punishable by death. Thus, the most valuable remedies were prescribed with apprehension of problems arising from their injudicious application. The mere existence of such a law in ancient Hindustan evidences the paucity in medical knowledge among the people who framed it, as this law presupposes that every poison has an appropriate and specific antidote, just as every disease was believed, down almost to the nineteenth century, to have a specific remedy.

Medicine in Ancient China

Europe is indebted to the nations of the East for so much of her early knowledge. For the Chinese, the study of medicine was synchronous with the foundation of their empire. The medical code that serves as a guide to the practice of the Chinese physicians was the production of a writer named Huang di, who is thought to have lived more than two thousand

[3] An error in translating the Old Testament for the King James Bible contributed to the false view that medicine was a distinct profession in Egypt. Refer to chapter 1 of the original text for a more elaborate discussion of this translational error.

years before Christ. However, the Chinese possess numberless works of great antiquity and deep erudition in which the opinions, practices, and prescriptions of the respective authors are clearly and systematically arranged.[4] The characteristic hypothesis found in most of these works is that every disease acts successively upon the heart, liver, lungs, and veins. The crisis of the disease is occasioned by its transition from one of these organs to another, and it is critical to be able to distinguish the proper stage at which one should treat the disease with direct remedies, arrest its progress, or merely change its course in order to weaken it; and to determine when a crisis is to be hastened or retarded, or when the cure should be left wholly in the hands of Nature.

The Chinese contend that vital heat and radical moisture constitute the two natural principles of life, and the blood and spirits are the vehicles for these two principles. They place the seat of radical moisture in the heart, lungs, liver, and veins. On the contrary, they place vital heat in the intestines, which they believe are six in number. According to Du Halde, the Chinese believe that the components of the body each have a proper temperament suited to their figure, situation, and particular uses. In short, the basis of their medical practice rests upon the hypothesis that a general harmony exists among all parts of the body.

Although the practices and theories of the ancient Chinese physicians had little claim to rationality in general, in some instances the Chinese appear to have been ahead of the Europeans in the field of discovery. The circulation of the blood is said to have been understood by the Chinese long before the discovery made by Harvey[5], and their determination of prognosis for disease is founded chiefly upon an accurate knowledge of the pulse's variations. One of their most ancient medical writers, who flourished about 200 years before Christ, composed an elaborate treatise on the pulse, which still serves them as a guide and is held in high regard. In feeling the pulse, they observe many minute details, dedicate considerable time to examining the beats, and compare their difference, by means of which they allege to discover the seat of disease, and to be able to tell the patient not only how and when his complaint will terminate, but even the spot where he feels pain, without asking the patient a single question. They do not, however, rely solely on the pulse to form their diagnoses and prognoses, and examine likewise the color of the face and eyes, the state of the tongue, nostrils and ears, and the quality of the voice.

Inoculation for Smallpox in China

The Chinese tend to trace the origin of smallpox to a period 1,212 years before Christ, or 20 years before the Trojan War. If this were really the case, it would be strange for us to admit that we are indebted for our earliest information about this disease to Ahrun or Aaron, a presbyter of Alexandria, who flourished about the time of Mahomed, or greater than 600 years after Christ. As history informs us, Pythagoras, who lived little less than 600 years before Christ, travelled into the regions of the East in search of knowledge. During his travels, he must have heard something about smallpox had it existed in China, which is so close to the

[4] A general description of China, translated from the French of the Abbe Grozier. London, 1788. 2 vols. 8vo. Vol. 2, page 487.

[5] Grosier, l. c. p. 482. Grosier's words, for which, however, he does not cite any authority, are, "it appears certain that the Chinese were acquainted with the circulation of the blood long before any of the nations of Europe."

countries that he visited. However, we have the strongest evidence in the extant medical works of the Greeks of their utter ignorance of smallpox, concerning which consequently neither Pythagoras nor any other oriental traveler could have brought back the slightest information.

Inoculation as prophylaxis against smallpox is also said to have been practiced in China long before it was introduced to Europe. A comment by Grosier regarding the Chinese's view of vaccination for smallpox must be noted. Grosier's comment is all the more important for being written many years before Jenner's discovery of the prophylactic powers of viral vaccination. Grosier's comment is consequently free from all suspicion of being biased towards one side or another with regards to the debate over the viral vaccine's efficiency in fortifying the human body against smallpox. Grosier's comment is that the Chinese place "less confidence" in the practice of inoculation "than the Europeans, and for this reason, because the former are convinced, by numberless instances, that it does not prevent a return of the smallpox when it becomes epidemical."[6]

This fact appears to have been overlooked by European practitioners until the controversy respecting the prophylactic powers of both the vaccine and the smallpox virus directed their attention to the subject. In the attempt to prove that the vaccine is inefficient in rendering the human body insusceptible to smallpox, it was ascertained almost beyond doubt not only that the smallpox vaccine is insufficient for protecting the patient from any subsequent attack, as the experience of the Chinese had taught them, but also that the disorder, even when acquired naturally, does not always protect the patient from recurrence of the disease. The Chinese tend to discriminate no less than 40 different species of smallpox and regulate their treatment not only by the nature of the climate and season, but also by the age and habit of the patient.

Judicial Medicine and the Chinese Materia Medica

Among other instances of superior sagacity to which they claim, the Chinese claim proficiency in determining the cause of an individual's death, and the process by which they come to such conclusions is essential to the administration of justice. Most remarkable is the method by which the Chinese discover whether a man found dead by strangulation had strangled himself or had been strangled by others; whether in the case of a body being found in water, death preceded or followed its immersion; and whether a person's death resulted from natural causes or felonious violence, and this last conclusion the Chinese claim to be able to reach after the body has been interred for some time and decomposition of the organs has begun and even after only the dry skeleton remains.

In suspicious cases, the body is exhumed and carefully washed with vinegar. Next, a large fire is kindled in a pit measuring six feet in length, three feet in width, and as much in depth. The fire is progressively augmented until the surrounding earth becomes intensely heated. When the fire is removed, a large quantity of liquor fermented from rice and honey is poured in, and the mouth of the pit is covered with a hurdle, upon which the body is stretched out at full length. A cloth, supported in the form of an arch, is thrown over both the body and the hurdle in order to confine the vapor arising from the liquor and direct its action to every part of the body. Two hours later, the cloth is removed, and the body is inspected. If any

[6] Grosier, l. c. p. 489.

blows have been inflicted upon the body, their marks will appear distinctly. Even after total decomposition, when nothing but the naked skeleton survives, this process is sufficient to render visible the marks of any blows that may have been received and were sufficient to cause death, even though no bones were fractured.[7]

Such is the skill to which the Chinese lay claim in the field of medical jurisprudence. However, when we examine their medical writings, we find them replete with the grossest absurdities and destitute, or nearly so, of all real information. The numerous schools of medicine and astrology in which their youth are said to have been formerly instructed live only in the floating visions of recollection and the legends of tradition, if ever they actually existed. Like their works on therapy and pathology, the Chinese materia medica exhibits a mixture of sometimes useless simples, worthless remedies, and superstitious prescriptions. However, we discover some articles worthy of acknowledgement and for our knowledge of which we are indebted to the Chinese, such as the ginseng plant (*Panax quinquefolius*). The Chinese have over 77 different preparations of ginseng root, which was in such demand among them as to be worth nearly its weight in gold.

Progress of Medicine among the Greeks

The history of medicine in the earlier ages of Greece is enveloped in mystery and fable. History tells us that the greater portion, if not the whole of the philosophy and erudition of the Greek sages, was borrowed from the magi of Egypt. Pythagoras, who lived about six centuries after Asclepius, sought a degree of knowledge beyond what Egypt could supply and traversed the vast peninsula of India, and brought back the doctrine of the transmigration of souls, the mysterious powers of numbers, and other metaphysical dogma of the Brahmins[8]; but, given the state of medical knowledge in Hindustan, as previously discussed, travelers could not have acquired from those regions any knowledge that would have been substantial additions to the stock of medical information previously existing either in Egypt or in Greece. Hence, the medical traditions of the Greeks probably were wholly derived from the schools of the Egyptians, and thus the respective medical mythologies of Greece and Egypt are essentially the same.

A notable figure in Greek mythology is Orpheus, the son of Apollo and Calliope, who is described as having inherited from his mother the powers of melody that immortalized his name, and from his father all the skill in the art of the healing that was the distinguishing characteristic of the Egyptian character Thouth. However, according to some chronologists, Orpheus is not entitled to the praise that Greek mythology has awarded him of being the first who instructed his countrymen in the science of medicine. He is said to have been preceded by Melampus, the son of Amythaon and Dorippe, who gave the most convincing proof of his acquaintance not only with the art of healing but also with the virtues of plants through his curing the daughters of Proetus, king of Argos, when laboring under the combined influence of melancholy, madness, and leprosy. Melampus appears to have united the art of prophesying with the practice of medicine, and is commemorated through the name of

[7] Grosier, l. c. p. 490.

[8] For a review of the history of Indian philosophy and fundamental principles of Indian religious traditions, the authors recommend *Hinduism, Its Historical Development* by Troy Wilson Organ.

Melampodium[9] given to Hellebore, the plant with which we are told he successfully combated the insanity of the daughters of Proetus.

Chiron

The next prominent practitioner recorded in the medical annals of ancient Greece is Chiron, the fifth son of Saturn and Phylira, whose skill in and fondness for horsemanship earned him the reputation of being a centaur. This hero of mythological romance appears to have flourished between the periods of the celebrated Argonautic expedition and the siege of Troy, the former having taken place about 1,234 years and the latter 1,192 years before Christ. Chiron must have lived to a very advanced age if, as we are informed, he instructed the leaders of both those expeditions in medical and surgical practice. According to Xenophon, Chiron had numerous pupils, among them Amphiaraus, the son of Oecleus, who was carried against his will to the siege of Thebes and perished in an earthquake on the day of his arrival as he had himself foreseen; Castor, a Greek physician who is said to have been the first who used pepperwort (*Lepidium*) to manage epilepsy; Asclepius, the reputed son of Apollo and the nymph Coronis, who, with his two sons Machaon and Podalyrius, has been immortalized by Homer in the *Iliad* as a skillful surgeon; and Hippolytus, the son of Theseus whom Asclepius is fabled to have restored to life at the intercession of Diana.

Asclepius

Among all of Chiron's pupils, Asclepius (Figure 1) and his two sons became most renowned for their professional skill and successful practice of medicine. Asclepius has been referred to as the founder of the science of medicine. He may have been born in Thessaly, and we can imagine that he exercised the power of a sovereign over that territory. Notwithstanding the fables interwoven with his history, we can collect enough from it to believe that his talents must have been of the very first order and that his knowledge must have appeared supernatural. According to Cicero, no fewer than three individuals were distinguished by the name of Asclepius, and all were celebrated for their proficiency in medicine and surgery, the son of Apollo and Coronis to whom is ascribed the invention of the probe and bandages for wounds and who was worshipped chiefly in Arcadia and may be regarded as the one so justly famed in medical history; the brother of Mercury; and the son of Aristippus and Arsinoe, who first introduced the use of purgatives and the practice of tooth-drawing. However, the practice of purging, if not synchronous with the dawn of medical practice, existed long before Asclepius, as Melampus cured the daughters of Proetus by using

[9] Melampodium is a name applied in modern botany to a very distinct genus from the Hellebore of the ancients, being plants of the new world, belonging to the Linnean class and order of Syngenesia Polygamia Necessaria, and not possessing the peculiar cathartic properties of the celebrated Black and White Hellebores of the ancients, the former of which is believed to have been the same with the Ilelleborus niger, or Christmas Rose of the modern botanists, a plant belonging to the class Polyandria, and order Polygynia, flowering from January to March, and the root of which is drastically cathartic; while the latter is supposed to be the Veratrum album or White Hellebore of the moderns, a plant belonging to the class Polygamia and order Monoecia, the root of which possesses similar properties.

the purgative Hellebore. Upon his death, Asclepius was glorified as a god, and temples were built in which he was worshipped. The most splendid of these temples was in Epidaurus, where every five years games took place in honor of him.

Figure 1. Asclepius (Image courtesy of Reynolds Historical Library, the University of Alabama at Birmingham).

No certain portions of the doctrines or practice of Chiron and Asclepius have been handed down to us, and the descriptions of their exploits in the field of medicine that we find in the writings of the poets do not afford us any correct means by which to estimate their real merits. During the five to six centuries that elapsed between the siege of Troy and the age of Hippocrates, we find the various branches of medicine occupied chiefly by the priests. A spirit of rivalry among the priests, inspired by their devotion to their respective deities, promoted competition conducive to the advancement of medical science, as it compelled the priests to call in the aid of observation to the cause of superstition, and to exercise their natural talents to accomplish those cures, the credit for which they gave to their gods. The temples of Asclepius were especially known for the success of their practice, and, being served chiefly by the descendants of Asclepius, may be regarded as forming the most faithful depositaries of his doctrines and the best schools of his practice.

In order to add to the celebrity of their temples and to accumulate important practical information, the following practice was instilled in patients as a duty of religious gratitude that every patient was bound to fulfill upon recovery, a representation of the appearance of his disease or an accurate model of the part affected, accompanied by a tablet[10] describing the symptoms and recording the cure, would be displayed in the temple of the divinity whom the patient credited for his or her recovery. The accumulation of these medical trophies signified the reputation and success of the temple, and contributed to the improvement of the priests' practice of medicine by providing them with instructive facts and observations. Thus, these temples were progressively converted into schools of medicine. In addition, the foundation was established for the great revolution in medicine that Hippocrates first initiated, and that detached medicine altogether from theology and elevated it in time to the dignity of a rational science.

These temples, or schools of medicine, did not differ from each other more widely in the measure of their fame than in the nature of their doctrines and the methods of their practice. Empiricism, in its strictest sense, was the distinguishing feature of the School of Cnidus in Asia Minor. This school was a seminary that produced Euryphon, the author of the Cnidian Sentences, and Ctesias, fragments of whose history of Persia have been preserved among the writings of Photius. In contrast, the School of Cos, in which Hippocrates received his first instructions in healing, partook more largely of the dogmatic character and laid the greatest stress upon acquiring knowledge of the diagnostic symptoms, remote and proximate causes, and systematic treatment of various human diseases. For a period after the death of Asclepius, medical instruction at these schools was reserved exclusively for his descendants, the Asclepiadæ. Eventually, other students were admitted on the condition that they agree to conform to the rules of the Asclepiadæ.

Pythagoras, Alcmaeon, and Empedocles

Sages notable for their talents and achievements had already directed their attention to medical research. Among these contributors, none deserves more credit than Pythagoras (Figure 2), who was born on the island of Samos 600 years before Christ. Pythagoras left Samos while it was under the dominion of Polycrates. His search for knowledge led him first to Egypt. We have previously described his travels to India, yet we have no evidence to answer the question of whether he journeyed into China beyond that which may be gleaned from portions of his doctrines, none of which appears to support the idea that his travels in the East went beyond India. Upon returning to Europe, he settled in Crotona, a small town of Magna Grecia in Italy, where he founded a school about the time, as Cicero informs us, of Tarquinius Superbus. At this school, he taught his pupils the theory, if not the practice, of medicine, promoted the study of the animal economy, and introduced a regular system of dietetics. By following the teachings he brought back from the East, Pythagoras exposed himself to ridicule, such as for attempting to account for everything through the power of numbers. His system of therapeutics hardly rose beyond the level of that taught in the temples. Although he fell short of the merit of Hippocrates, Pythagoras established rules for the regulation of diet and regimen, and this accomplishment reflects a rare union of natural talent and profound observation.

---

[10] Specimens of a few of these votive tablets have been preserved for the gratification of the curious by the industry of a German writer of the name of Grüter, among whose works they may be found.

Figure 2. Pythagoras (Image courtesy of the National Library of Medicine).

It does not appear that Pythagoras mustered the courage to challenge the prejudices of his age by dissecting the human body. However, comparative anatomy, or the dissection of the bodies of animals, seems to have been studied often among his pupils. One pupil, named Alcmæon, is represented by the testimonies of Aristotle, Diogenes, and Plutarch as having acquired considerable skill in comparative anatomy. Chalcidius, in his commentaries on the Timæus of Plato, asserts that he was the first who dissected animals in order to discover the internal structure of their bodies. Alcmæon's opinions appear to have been somewhat paradoxical. He claimed that respiration in goats was partly carried on through their ears. He attributed health to accurate adjustment of heat and dryness, coldness and humidity, bitterness and sweetness, and other qualities, and he considered the derangement of this balance and the excess of any one of these qualities to be productive of disease. He said that hearing resulted from the concave form of the interior of the ear, with all hollow places resounding when any noise enters them. He identified the brain as the seat of the soul and thought that the brain receives inhaled odors and thereby produces the sense of smell. He imagined that the tongue was able to discriminate tastes through its humidity, moderate heat, and softness. According to Plutarch, Alcmaeon argued that sleeping resulted from the blood returning to the larger vessels and that waking resulted from the blood redistributing throughout the system. These opinions on physiology entertained by Alcmaeon do not suggest a very accurate comprehension of the internal structure of the human body on his part and support our refusal to endorse the claim that he practiced human anatomy. There are the strongest grounds for believing that the dissection of the human body was not practiced, or at least not publicly taught, before the time of Herophilus, who flourished during the reign of Ptolemy Soter a little more than 300 years before Christ and nearly a century later than Hippocrates. We may reasonably conclude that if Alcmaeon possessed the courage to overcome the superstitious prejudices of his age, his daring did not lead him to venture farther than dissecting apes, whose external form resembles the outline of the human body and thus might have led him to expect a similar correspondence between the respective internal structures of humans and apes. Thus, Alcmaeon may have acquired the reputation of an anatomist from having been seen in the act of dissecting an ape.

Empedocles (Figure 3), a poet and philosopher of Agrigentum, a town of Sicily, who flourished about 578 years before Christ, was also a pupil of Pythagoras, and entertained, as we learn from Galen, some paradoxical opinions concerning the structure and functions of the various parts of animals. The fact that Empedocles contended that respiration commences before birth indicates that he was little acquainted with the true situation of the fetus in utero or the mode by which its circulation is connected with the vascular system of the mother through the umbilical cord and placenta. His notions respecting the internal structure of the ear and the manner in which hearing is produced were equally vague and erroneous. Empedocles contended that hearing resulted from the impulse of the air striking against the interior of the ear, which he fancied to be convoluted like a shell and fixed to the most elevated part of the body like a small bell, capable of sensing every undulation of the air entering from without. In conformity with the prevailing opinions of the day, he taught that flesh was formed of equal portions of the four elements. He believed that two parts of water, one of fire, and as much of earth went to form the nerves, the extremities of which, cooled and indurated by the surrounding air, formed the nails, and that the bones consisted of equal parts of earth and water, or at least that these two elements preponderated in the formation of bones. He viewed sweat and tears as the thinner parts of the blood, and wrote a book in Greek on the nature of things, in which he compares the seeds of plants to the eggs of animals, both of them dropping when they have become mature.

Despite his flawed analysis of physiology and apparent ignorance of the structure of the human body, Empedocles possessed a mind superior to those of many of his contemporaries and a power of discrimination that enabled him, upon two occasions, to perform essential services for his fellow men. The first instance occurred on the occasion of a plague afflicting the inhabitants of Agrigentum, which he attributed to the entrance of noxious fumes into the city through openings in the nearby mountains. By closing the openings, he succeeded in combating the disease and preventing its recurrence. The second instance occurred on the occasion of a plague afflicting the inhabitants of Selinus, a town of Cilicia, which was due to stagnation of the river surrounding the town's walls. He ordered the streams of two neighboring rivers to be turned towards the main river, thereby washing away the filth in the river bed and restoring health to the town's inhabitants.[11]

[11] Diog. Laert. Lib. viii. Sogm. 70.

Figure 3. Empedocles (Image courtesy of the National Library of Medicine).

Commencement of the Practice of Medicine as a Distinct Profession

The practice of medicine had been chiefly monopolized by the priests of the several temples that had acquired a reputation for successful treatment of disease, but their practice and instructions were equally confined to the precincts of their respective shrines. The school founded by Pythagoras at Crotona appears to have caused some change in this respect, and to have partially at least wrested the office of medical instruction from the grasp of the priests. However, the doctrines of Crotona were more speculative than practical, and more theoretical than experimental. The philosophers of this school for the most part spent more time thinking about disease than treating disease, and studied symptoms and cures while in their homes rather than in the chambers of sick patients. Society was still lacking in persons who pursued the art of healing as a distinct and practical profession that was not connected to the priesthood. About this time, an important change essential for the improvement of the science and for the welfare of patients took place. A group of men, some of them pupils at the school in Crotona, pursued the study of medicine with a view to practical utility and practiced medicine as a distinct and independent profession, offering their services to all who required their aid. These practitioners travelled frequently from place to place as their services were required and visited their patients at their own houses, unlike the priests of the temples and the philosophers of the schools, who confined their practice to one particular spot. Thus, these physicians came to be known as travelling or itinerant physicians, some of whom attained considerable repute.

Democetes and Acron

Democetes of Crotona, an itinerant physician, was retained as a family physician by Polycrates, king of Samos, thus evidencing that many of the aforementioned practitioners, whether stationary or travelling, were men of reputed skill and held in high esteem. Democetes, whose practice included surgery, was held captive in Persia, and he gained additional fame by curing Darius of a sprain that had baffled Egyptian practitioners. Furthermore, he successfully treated a painful tumor of the breast that Darius's wife Atossa, the daughter of Cyrus the Great, had suffered for a long time without the slightest relief. Acron of Agrigentum, a contemporary and rival of Empedocles, was also an eminent itinerant physician, and the author of several medical works written in the Doric dialect. During the plague that depopulated Athens about 473 years before Christ, Acron's professional services were critical in arresting the spread of the contagion and diminishing the number of deaths from the epidemic, which he accomplished, as Plutarch informs us, by recommending large fires to be kindled in the streets and by introducing the practice of fumigation.

The Gymnastic Schools of Medicine

About this time, schools for gymnastic exercises became extensively established throughout Greece and her colonial settlements. These schools contributed to the spread of medical knowledge and the improvement of medical practice, and each school in fact became a distinct school of medicine. This trend gave birth eventually to the Gymnastic school of medicine, which was under the auspices of Iccus, a practitioner of Tarentum in Calabria, and Herodicus of Selymbria or Selivria, a city in Romania, not far from the place where Constantine the Great founded Constantinople. At these schools for athletic exercises, where a strict attention to the diet and regimen of the pupils was essential for their proficiency in these manly sports, the superintendence of this department was the responsibility of the keeper of the seminary for these exercises, while the persons who presided over the exercises were required to have a thorough knowledge of diseases and their symptoms and treatments so as to be able to apply timely remedies. The surgical responsibilities of the department were assigned to the persons employed to anoint the youths in preparation for the commencement of their exercises. These anointers likely encountered individuals with fractures, luxations, and a variety of other injuries and performed phlebotomies almost daily.

Chapter 2

Reformation of Medicine by Hippocrates

The Opinions and Theories of Hippocrates

We have now arrived at the second stage in the history of medicine, when a revolution was accomplished by the genius of Hippocrates (460-361 B.C.). Hippocrates (Figure 4) was born on the island of Cos. His main objectives were to emancipate medicine from false canons, and to establish its fundamental principles upon a solid and rational basis. With this view, he combated the doctrine of the celestial origin of disease, which could prove highly detrimental to the patient by paralyzing the efforts of the physician. On this subject he expressly says that "no disease comes from the gods, one more than another, each acknowledging its own natural and manifest cause." In his treatise *De Morbo Sacro*, he teaches men how to do away with false religious notions.[1] Our chief aim, he tells us, should be to learn the true properties of things by actual experiments, patient investigation, and careful deduction. Hippocrates ascribes all the phenomena of life and health to the operations of a fundamental principle, which he terms "nature". He conceives that nature is sufficient for the preservation of every animal and acts by attracting what is good and rejecting what is bad, thereby forming the foundation of the theory of depuration, concoction, and crisis, upon which he and his followers practiced.

Hippocrates has written on the subject of diet, making it the principal fulcrum of his practice. He assigns various causes for disease, such as the humors of the body, errors of diet, and impurities of the air. He distinguishes disorders into four separate stages, namely their beginning, progression, peak, and decline, with the most considerable change being in the third stage, which determines the fate of the patient and in general constitutes the crisis. This process proceeds from an effort of nature to restore the humors to their natural state, and this is affected by concoction, the time required for which varies with circumstances. When the duration of the complaint exceeds 60 days, he considers the case as chronic. His chief merit

---

[1] The attempt of Hippocrates, as Miller has indicated, to extinguish "the popular belief in the supernatural causation" of the diseases is "one of the earliest and most significant expressions of the self-consciously rational spirit in early Greek medical thought." See, Miller HW. The Concept of the Divine in De Morbo Sacro. Transactions and Proceedings of the American Philological Association. 1953; 84, 1-15.

was his efforts in watching and accurately recording the minutest aspects of the disease, and describing every accident preceding or accompanying the attack and everything that tended to alleviate or aggravate its severity. By such careful observation, Hippocrates was able to predict the nature of the coming illness and to anticipate the time and manner of its termination.

Figure 4. Hippocrates, 460-375 B.C. (Image courtesy of Reynolds Historical Library, the University of Alabama at Birmingham).

Hippocrates' judgment in disease was founded on the appearance, posture, urine, feces, expectoration, sweat, and pulse of the patient. The means he employed for the preservation of health and the cure of disease were chiefly diet and regimen. As a means of preserving health, he recommended the use of emetics once or twice a month during the winter and spring seasons. Among his principles for the cure of disease are that "contraries are remedies for each other"; thus, evacuation is the remedy for repletion, and repletion for depletion. Everything in excess is repugnant to nature.

In acute disorders, especially in febrile illnesses, Hippocrates preferred the consumption of liquids to that of solids. He disapproved of the sick eating more frequently than those in health. During the paroxysm of fever, he prohibited food altogether. He did not approve of the rigid system of abstinence with which the practitioners of his day treated every complaint. With respect to drinks, he disliked plain water, substituting for it various cooling and acidic beverages, and not absolutely forbidding wine. In complaints of a chronic nature, he made great use of milk and whey. Bathing was among his favorite remedies. In the use of the bath, he cautions against catching a cold and prohibits bathing immediately after eating or drinking.

We should not bathe, he says, when nausea, loss of appetite, or a disposition to bleeding from the nose prevails. To those who are accustomed to it, he allows bathing twice every day. In chronic cases, he approved of exercise, although he thought it mildly contraindicated in acute disorders. He did not, however, recommend constant confinement to bed.

When diet and exercise proved insufficient to conquer disease, he used more active remedies, among which purging was one of the first. He imagined that purgatives operated on the malignant humors and discharged them from the body. The ancients appear to have had little acquaintance with those prokinetics, for the introduction of which we are indebted to the Arabs at a considerably later period. In the consumption of these remedies, Hippocrates was cautious by never using them during the summer months or administering them to pregnant women, in whom they may have produced abortion, and only seldom giving them to the elderly or infants. His principal rule for purging was to remove concocted humors, not those that are crude, thereby avoiding purging in the beginning of a disorder or between the first and fourth days. Believing that purgatives operate by some peculiar attraction for the humors, he concluded that each humor has its specific purgative. When he wanted to recall the humors from the innermost recesses of the body, he used more potent emetics such as the White Hellebore, which he gave primarily to insane and melancholic patients.

He made frequent use of phlebotomy, not merely to achieve simple evacuation, but also to recall the blood when taking a wrong course and to procure free motion of the blood and spirits. In complaints seated above the liver, he took blood from the arm, but in those situated below the liver, he opened the veins of the foot, ankle, or posterior thigh. He sometimes carried bleeding to an extreme even in chronic diseases. He also bled patients with edema and enlargement of the spleen, performing the procedure sometimes for so long that it caused syncope. Occasionally, he employed cupping and scarifying. When purging and bleeding failed to afford the relief expected, he called in the aid of diaphoretics and diuretics. He also frequently employed gargling with quinsy and fumigations, especially in female patients. He made considerable use of oils and ointments, medicated plasters, and eyewash.

Hippocrates is said to have been the first author who treated anatomy as a science. His knowledge of the bones of the human frame is said to have been realized by the brazen model of a skeleton that he hung up in the temple of Apollo[2] as testimony to his diligence and skill as well as for the instruction of future generations. The best informed writers have doubted that Hippocrates extended his dissections beyond the limits of comparative anatomy or dissection of animals. Had he really dissected human bodies, it is unlikely that he would have confounded the nerves with the tendons and ligaments. It is also unlikely that the information that we may presume he communicated to his pupils could have been so effectually lost that Diocles, Praxagoras, and others more than a century later should have been such unskillful anatomists. Diocles[3] is believed by many to have lived almost immediately after the time of Hippocrates and to have been a contemporary of Plato, and he must have seen and conversed with some of his immediate descendants whom we doubt were ignorant of his anatomic

---

[2] The temple of Apollo at Delphi in Greece was built in the eighth century B.C. See, International Dictionary of Historic Places, Southern Europe. Chicago, Fitzroy Dearborn, 1995.

[3] Diocles, for his fame and services, has been called the second Hippocrates. See, The London encyclopaedia, Volume 14. 1839, p 15.

knowledge. Praxagoras[4] was a descendant of Asclepius, and thus must have been instructed in anatomy.

Had the works of Hippocrates been handed down to us uncontaminated, we might have been led to suspect that he had some crude notions about the circulation of the blood. Despite his deficits, Hippocrates (Figure 5) was indisputably the most surprising man of his age, and his opinions upon many points retain their full authority even today. He died in Larissa, Thessaly, around 361 B.C. at the age of 101 years.

Figure 5. Hippocrates seated, D'Artaxerces pouring money at his feet. "*Hippocrate Refusant Les Presents D'Artaxerces*" by French painter Anne-Louis Girodet de Roussy-Trioson (Image courtesy of the National Library of Medicine).

Democritus of Abdera

Contemporary with Hippocrates was Democritus of Abdera[5], whose excessive fondness for anatomical research subjected him to the accusation of insanity. Hippocrates, who was sent for to visit him, found him engaged in the dissection of animals with a view to discover the cause of disease, which he imagined came from the bile. Hippocrates reported to his

4 Praxagoras of Cos taught Herophilus of Alexandria in anatomy. See, Wiltse LL, Pait TG. Herophilus of Alexandria (325-255 B.C.), The father of anatomy. Spine (Phila Pa 1976) 1998; 23, 1904-14.

5 Democritus had travelled to the East visiting several philosophers. See, Tsoucalas G, et al. Democritus (460-370 BC) on embryology, anatomy and pediatrics, the unknown aspects of the Greek atomic scientist. Ital J Anat Embryol 2012; 117, 199-204. His "atomic hypothesis", stating that things are composed of small and invisible particles, is remarkable.

employers not only that Democritus was in the full possession of his senses, but also that he was the wisest of men. From this anecdote, it is evident that the study pursued by Democritus (Figure 6) was nothing more than comparative anatomy, from which physicians were accustomed to reason by analogy the structure of man.

Figure 6. Democritus drawn by A. Day (Image courtesy of the National Library of Medicine).

The Dogmatic School

Hippocrates had two sons, Thessalus and Draco. These two individuals, along with Hippocrates' son-in-law Polybus,[6] were the true founders of the Dogmatic school and the

[6] Polybus of Cos (approximately 400 B.C.) is credited with writing The Nature of Man, which advanced the theory of humoralism. Corpus Aristotelicum quoted Polybus as mentioning a unique theory regarding the body vasculature being composed of four pairs of blood vessels originating from the head. A numerological mysticism might have been the motive for both humoralism and Polybus's theory of the vasculature. See, Shoja MM, et al. Wrong theories on the origin of blood vessels, Polybus and De Natura Hominis. Int J Cardiol 2008; 126, 313-5.

authors of many works that have been handed down to us in Hippocrates' name. This school, misapplying the mystical speculations of Platonic philosophy to the study of medicine, adopted the principle that "where observation failed reason might suffice", thus rejecting the leading feature of the system that produced such profound results under Hippocrates and deviating from the only path that could conduct them to truth. So the followers of this school neglected or rejected with disdain the patient observation of nature upon which Hippocrates insisted. They deceived themselves into thinking that they had laid the foundation of an incontestable system, while the accumulation of facts from experience and observation was yet in its earliest infancy.

Eudoxus, Chrysippus, and Diocles

About this time, Eudoxus of Cnidos[7] and his pupil Chrysippus introduced the Pythagorean system, accompanied by a modification of Egyptian practice. Chrysippus had an aversion both to purgatives and bloodletting, but a strong inclination for a vegetable diet. Contemporary with Plato, according to some, or about half a century later under the reign of Antigonus lived Diocles of Carystius, who was second only to his great predecessor Hippocrates in point of reputation. He is said to have devoted much of his time and attention to comparative anatomy and to have corrected many of the errors of his predecessors. Even Galen informs us that Diocles, whom he criticized for his ignorance of anatomy, was the first to write a manual on the subject of the dissection of cadavers. It is difficult to reconcile these conflicting statements. Diocles is said, like Eudoxus and Chrysippus, to have blended the mystical doctrines of Pythagoras with the practice of medicine, and to have ascribed wonderful properties to the number seven and its various combinations. He developed a surgical instrument for extracting arrows that resulted in less suffering for the patient.

Praxagoras and Menecrates

Praxagoras of Cos is supposed to have lived nearly at the same time as Diocles and to be one of the descendants of Asclepius. He is said to have distinguished himself, contrary to the assertion of Galen, by his anatomical studies and by being the first to distinguish between arteries and veins. Among other doctrines taught by Praxagoras was the one that refers the origin of fevers to an inflammation of the vena cavae. He demonstrated the absence of cotyledons in the human uterus, explained the principle of the pulse, and reduced the humoral pathology to a more regular system. It thus would appear, contrary to the opinion of Galen, that he was an able anatomist and that he did dissect human bodies. On this latter point, however, we lack direct evidence. Unlike Chrysippus, the pupil of Eudoxus, he was a strenuous advocate for bleeding, and he made great use of emetics, which he used extensively in cases of ileus. As a surgeon, he was equally bold and skillful, not hesitating to remove accumulated feces in cases of violent feculent emesis by opening the cavity of the abdomen

[7] Eudoxus was born in Cnidos, Turkey, in approximately 400 B.C. He studied medicine under Philistion in Sicily. See, McEvilley T. The Shape of Ancient Thought, Comparative Studies in Greek and Indian Philosophies. Skyhorse Publishing Inc., 2012.

and opening the rectum. Nearly contemporary with him was the physician Menecrates of Syracuse, who flourished about 360 years B.C. and is memorable chiefly for his arrogance and for the punishment inflicted upon him by Philip of Macedon.

Aristotle, Theophrastus, Zeno, and Epicurus

Aristotle (Figure 7), the illustrious preceptor of Alexander the Great, flourished in the fourth century B.C. In human anatomy, he appears to have confused the functions of the arteries and the veins, the latter of which he imagined to receive the blood from the heart for distribution through the body. He denied the return of any blood to the heart. He gave the name "aorta" to the great artery of the body.

Figure 7. Aristotle (Image courtesy of the National Library of Medicine).

Aristotle, like his predecessor Hippocrates, confounded the nerves with the tendons and ligaments, and imagined that they proceeded from the heart. Indeed, he ascribed to the heart many of the nobler functions of the brain, maintaining that this last was void of blood and destitute of sensation, responsible only for balancing and correcting the heat of the heart. The liver, spleen, and kidneys served chiefly to support and fix the veins in their proper places. Equally remote from truth was his theory of digestion. His knowledge of the internal structure of the human body was derived solely from the dissection of animals, for he says that he had no actual knowledge of the internal parts of the human body. In his *Historia Animalium*, he mentions that the human saliva counters the bites of most serpents.

Theophrastus (Figure 8), one of Aristotle's pupils, also contributed much to the improvement of the science of medicine, especially in botany. He was the founder of the science of botany. Only two of his works have reached us, *Historia Plantarum* and *Moral Characters*. Zeno and Epicurus may be justly ranked among the promoters of medical improvement, not only from the additional impulse they gave to the general pursuit of knowledge, but also from the eagerness with which their principles were grasped by the Empirical school, in opposition to the Dogmatists, who adopted the opinions of the Stoics.

Figure 8. Theophrastus (top middle) in the title page of Latin edition of *De Historia Plantarum*, 1644 (Image courtesy of the National Library of Medicine).

Ptolemy Soter

With the death of Alexander the Great in 323 B.C., and following the division of his empire, Egypt fell to the lot of Ptolemy Soter. He promoted literature in all its various branches, encouraging men of science from all parts of the world to settle at Alexandria and form an academy or learned society, which was called the "Museum". For the use of this academy, around 304 B.C. he founded a splendid library. Ptolemy not only held out the most liberal encouragement to men of learning to visit his dominions, but also induced the merchants of Greece and Syria to make Alexandria a principal market for their commerce, and thus succeeded in rendering this city what its illustrious founder designed it to be. The establishment of this library, which swelled rapidly to upwards of 700,000 volumes along with the establishment of medical schools and hospitals, contributed significantly to the study of medicine. The result was not only that Alexandria far outstripped every other place in commercial wealth and literary reputation, but also that the school of medicine established there soon eclipsed every other in the number of its pupils and the celebrity of its professors and produced a rapid succession of eminent physicians, among whom Herophilus and Erasistratus (Figure 9) made the most important contributions to anatomical science early on.

Figure 9. Herophilus and Erasistratus (from a book by Lorenz Fries, 1532; Image courtesy of Wellcome Library, London).

Herophilus

Herophilus is generally believed to have been a native of Carthage and to have been born during the reign of Ptolemy Soter. He is said by some to have been a pupil of Praxagoras. He, along with Erasistratus, was accused of having dissected living subjects. Such accusations, however, appear to be nothing more than blind and mischievous superstitions. Galen himself declared that Herophilus was an able physician and a learned man, and stated that both Herophilus and Erasistratus conducted their anatomical researches chiefly at Alexandria, and probably under the protection of the Ptolemies.

The discovery of the true nerves, which were unknown both to Hippocrates and Aristotle, is ascribed to Herophilus, who arranged them, according to Rufus of Ephesus, in three divisions. The first division consisted of those real nerves that originate in the substance of the cerebrum, cerebellum, or spinal cord, and are the organs of motion and sensation throughout the system; he accordingly termed them "nerves of sensation and volition", those which communicated sensation and obeyed the commands of the will and could be traced either to the brain itself or to the spinal cord. What he arranged as nerves under the two remaining divisions were evidently nothing more than the tendons and ligaments, not actual nerves. He particularly described the optic nerves, which had, he maintained, a perceptible cavity not observable in any of the other nerves. He viewed the brain as the seat of the soul that resided in one of its ventricles. The cavity of the fourth ventricle of the brain he compared to that of a pen, and distinguished the point at which all the sinuses of the dura mater unite. He also first distinguished the retina and arachnoid layers of the eye. Herophilus noticed, under the term "palsy of the heart", a complaint of rare occurrence in his day, which may be what modern writers term "angina pectoris".

Erasistratus

Erasistratus is supposed to have been a contemporary of Herophilus and to have flourished about 300 B.C. during the reign of Seleucus, one of Alexander's captains who, upon the division of the empire, seized Syria for his share and established the Seleucid dynasty. He is admitted by all but Galen to have been an anatomist of the first skill and a practitioner of the first reputation. Like Herophilus, he conducted his anatomical studies at Alexandria under the protection of the Ptolemies and was accused of dissecting the living. He distinguished the nerves, as Rufus of Ephesus informs us, into two types, nerves that are the organs of sensation and nerves that are the organs of motion. He also described with considerable accuracy the valves that guard the several orifices of the heart. Despite his skill in anatomy and his great experience with dissection, he struggled under the popular misconception that prevailed respecting the functions of the arteries and believed them to be filled only with air. This error originated solely in his ignorance of the circulation for, as Galen informs us, he could not comprehend why there should be two distinct sets of vessels for the conveyance of the same fluid. In order to reconcile what appeared to him as an absurdity, he maintained that the vena cava was the reservoir of the blood and that the aorta was the recipient of the spirits, and that the blood did not enter the vessels in which the spirits flow, but noted that when this arrangement is disturbed, the blood forces its way into the arteries. The only use that he assigned to respiration was to supply the arteries with air.

Erasistratus studied the anatomy of the brain with considerable care and formed a tolerably correct idea of its function and importance. However, he does not appear to have suspected the slightest distinction of function between the cerebrum and the cerebellum. In his medical practice, Erasistratus, if we are to believe Galen, wholly banished the use of the lancet. We are informed by others that he was much more sparing of bleeding than other practitioners, for which he alleged the following reasons, the difficulty of success from not being always able to see the vein that requires to be opened, the danger of opening an artery by mistake, the difficulty of ascertaining the precise quantity of blood that should be taken, and the inutility of venesection to relieve inflammation of the arteries occasioned by a coagulation of the venous blood in their orifices. He also disliked purgatives, for which he substituted emetics and enemas.

His practice was on the whole the reverse of that of Herophilus, and he relied on diet and regimen aided occasionally by topical applications for the cure of disease. In the practice of surgery, he appears to have been a bold and successful operator, not hesitating in cases of hepatic tumors to make an incision at once through the skin and integuments into the cavity of the abdomen and apply his remedies directly to the part affected. However, he wholly disapproved of draining in the setting of anasarca, believing that the liver, being in an inflamed and schirrous state, would be more pressed upon and injured by the adjoining parts after the water was drawn off, and the operation would more likely cause the death of the patient. The art of healing around this time, as Celsus informs us, was divided into the three distinct branches of medicine, surgery, and pharmacy, each of which was pursued as a separate and independent profession, although it had been the custom for the same individual to practice all three before this time.

Empirics and Dogmatists

Serapion of Alexandria helped organize the Empirics into a distinct sect around 271 B.C. This sect actually dates its origin to a more remote period. From what we are able to collect of the practice of Serapion, the remedies he employed were the same as those in use among his predecessors, but administered without the slightest regard to the reasoning by which they were accustomed to account for their operation. The sect of the Empirics derives its name from its leading dogma, which held that experience is the only source of medical skill. Such experience might be fortuitous when resulting from personal observation, or acquired by design from studying cases recorded by others. The knowledge of a disease was to be obtained by comparing its symptoms with those of others already described, and the manner of treating a new complaint or one whose symptoms differed from symptoms already on record was to be determined by carefully observing which symptoms exhibited were in common with symptoms that were already known. The Empiric school held that observation was the ultimate source of knowledge.

While the attention of the Empirics was thus devoted to the investigation of symptoms, their Dogmatist rivals more rationally insisted on the importance of determining the causes from which those symptoms arose. It must be admitted that although the cause remains concealed, the effect cannot be removed with certainty. He who would attempt to cure a tubercular ulceration of the lungs merely by opiates to allay the most urgent symptoms would eventually lose his patient. The rationally empiric approach to such a case would be first to determine the latent or remote cause of the disease, which is the formation of tubercles; next, to determine the malignant factor, which is the successive inflammation of these tubercles; and lastly, to consider the natural indications of cure, the rational plan of restoring the diseased parts to a healthy state of action, and the safest method of alleviating the most distressing symptoms. Such a plan of action exhibits little in common with the practice or the theories of either the Empirics or the Dogmatists. The latter looked for latent causes of disease in imaginary changes among the elementary principles of our bodies. The cure of disease depends, as the Dogmatists contended, upon a due acquaintance with its cause, but they erred in passing over those causes that are the legitimate objects of investigation and maintaining useless beliefs about the ultimate arrangements of matter. The Empirics likewise erred in limiting their inquiries to an investigation of symptoms distinct from the underlying cause and skimming the surface, thereby ignoring the latent source of disease, which remained unaddressed. Such were the chief distinctions between these sects, but their practice in general differed minimally.

Heraclides, Nicander, Damocrates, and Herennius Philo

Heraclides of Tarentum, an avid student of the Empirical school, was particularly persistent in discovering the remote causes of disease, and his practice in treating cholera, tetanus, and various other dangerous maladies appears to have been of surprisingly sound judgment. His contributions to the *materia medica* of his day were copious, and his treatise on pharmacy derived additional value by being founded upon his personal observation. His skill in surgery has also been well spoken of, and his merits elicited commendation from Galen, who was in general sparing of his praise to any but Hippocrates. After the death of Heraclides, the research of medical men became focused on finding antidotes to poisons.

Nicander of Colophon, a town of Asia Minor or Anatolia, flourished in the time of Attalus, King of Pergamus, around 135 years B.C. He enjoyed a high reputation not only for his skill in healing, but also for his general attainments in grammar and poetry. The only works of his that have survived and have reached down to our days display an acquaintance with natural history. Damocrates, who lived around the same time, introduced a variety of the most complex prescriptions, such as mithridatium or Damocratic confection, a formula for the preparation of which may be found in the pharmacopoeia of the Royal College of Physicians as late as the year 1746. Herennius Philo, a native of Tarsus, invented a celebrated relaxant consisting of opium, euphorbium, and various aromatics that was formed into a confection. In the pharmacopoeia already mentioned, a somewhat similar preparation is detailed and known as philonium.

Medicine among the Romans

For the first 500 years of their existence as an independent people, the Romans appear to have lived without the skills of the physician. During that long period, a few simples constituted their entire *materia medica*. Around 187 B.C., the invasion of an epidemic that threatened their population compelled them to seek assistance beyond what their own defective knowledge supplied. The representatives of the Romans were dispatched to the temple erected in honor of Asclepius, the god of healing, by the inhabitants of Epidaurum to seek advice. The tradition relates that the oracle, in place of replying to their inquiry, gave the deputies one of the sacred serpents. Soon after the vessel reentered the Tiber on her return, the serpent sprang ashore upon an island just within the mouth of the river, on which the Romans afterwards erected a temple in honor of the mighty Asclepius. A second temple was soon erected by the Romans to the goddess Hygeia, the daughter of Asclepius who was worshipped by the Greeks as the goddess of health. Medicine in the earlier periods of Roman history and as late possibly as the time of the Emperors appears to have been practiced chiefly, if not exclusively, by slaves and freed men, upon whom the title of *Medici*, or physicians, was blindly bestowed. However, the importance of their services eventually obtained for them citizenship and privileges.

Archagathus and Menecrates

The first surgeon who voluntarily established himself in Rome appears to have been a Greek, as he was named Archagathus. The Senate not only offered him the freedom of the city, but even purchased a shop for him. The cruelty of his operations, however, which surpassed the comprehension of the Romans, eventually brought him into disrepute and resulted in his banishment from the city. A practitioner by the name of Menecrates flourished about the time of Tiberius, who was Emperor of Rome from 14 to 37 A.D. He is said to have invented the diachylon or lead plaster, which is still useful in surgical practice.

Asclepiades

Asclepiades, a native of Prusa in Bythinia, came to Rome around 96 B.C. after completing his studies at Alexandria and at Athens. He established himself in Rome as a teacher of rhetoric. He abandoned this job for the practice of medicine and soon succeeded in attracting notice and wealth. He claimed to have exclusive knowledge of the art of healing. He ridiculed Hippocrates for his patient observation of nature. Asclepiades denied nature to be anything but matter and motion, and thus inferred that Hippocrates did not understand his own use of the term "nature" when he spoke of nature as an intelligent being. He disputed the Hippocratic doctrine of crises, asserting that matter and motion were sufficient to explain all the phenomena of disease. As for critical days, finding that the crisis of fevers sometimes occurred on the sixth or eighth day, he rejected the whole doctrine as vain. He maintained that the crisis was never the result of any spontaneous operation of nature, but resulted wholly from the skill of the physician. He concluded his critique of the opinions and practices of Hippocrates and the whole body of ancient practitioners by stating that they merely watched how their patients died by focusing their attention on them and did not affect a cure. Adopting the atomic philosophy of Democritus and Epicurus, Asclepiades attempted to explain the various functions of the body and all the phenomena of health and disease by the doctrine of atoms and pores. He totally neglected and even despised anatomy.

Asclepiades employed wine in cases of lethargy with the diametrically opposite view of rousing and stimulating the patient. To these various methods of proceeding he added a most severe system of abstinence. He was no friend to emetics or cathartics, for the latter of which he usually substituted enemas. Among his favorite prescriptions was the use of cold water. He appears to have been indebted for his great popularity among the Romans chiefly to the free use of wine, which he prescribed when others withheld it.

Themison and the Methodic Sect

Asclepiades' pupil Themison, a native of Laodicea, founded the Methodic sect upon the leading principles of the system of Asclepiades, discarding, like his master, the study of remote causes and other opinions current among the practitioners of his day. He contended that a general acquaintance with the constituent principles of disease and due regard for those circumstances that are common to most complaints were all that was required in a physician. According to his philosophy, the science of medicine consisted of knowing what diseases have in common with each other. His practice corresponded with that of his preceptor and model, Asclepiades. Like other pretenders of his class, he seems to have enjoyed considerable reputation among the people. However, Juvenal, a Roman poet, has said the following about him, "How many sick in one short autumn fell, let Themison, their ruthless slayer, tell." Thus, some held him in contempt for his approach to medicine.

Themison was old when he founded the Methodic sect and did not live to see it arrive at perfection, but his successors, Soranus and Thessalus, did. A simplification of the theory and practice of medicine appears to have been their main objective. They were dissatisfied with the Dogmatic doctrine of supernatural causes and the Empiric method of relying on an investigation of symptoms alone. Thus, they endeavored to take an intermediate course and assumed, as the basis of their theory, certain conditions of the system that they supposed to be common to all diseases, forgetting that these conditions are often more obscure than even the superstitious causes of the Dogmatists. They classified complaints into two or at most three general classes, the first comprising those that result from stricture and the second are those that arise from relaxation. The last of these assume a mixed character. They also retained the divisions of chronic and acute, with the treatment for each being different. All complaints, if falling under one of his three great divisions, should be treated indiscriminately according to the method specified for disorders belonging to that particular division.

Thessalus

During the reign of Nero in the first century A.D., Themison was succeeded by Thessalus, a native of Trallis. He bestowed upon himself the ridiculous title of "Vanquisher of the Physicians", which he even had inscribed upon his tomb. No mountebank, as Pliny informs us, ever attracted more crowded audiences or gaping admirers than this pretender, who professed to make his pupils proficient in the whole science of medicine in the space of six months. He would obtain a multitude of scholars chiefly from the lowest ranks of society, and he took them on his visits to see patients. After six months, he conferred upon them the full privilege of commencing practice on their own account. Both Galen and Pliny represent his insolence and vanity as intolerable, and report that he was in the habit of ridiculing and despising all the regular physicians.

Soranus and Cœlius Aurelianus

The reign of Trajan appears to have been favorable for the progress of medicine. Both Soranus and Cœlius Aurelianus belonged to the school of the Methodists. Soranus, a native of Ephesus, after having completed his studies at Alexandria, established himself in practice in Rome and studied anatomy.

He wrote a book on the female organs of reproduction. He has also left us a life of Hippocrates that contains much valuable information. Soranus also furnishes us with evidence against the story about how Hippocrates arrested the progress of the plague in Athens by means of fire, this error having been propagated even by Galen. Soranus observes that Hippocrates predicted the pestilence and took the necessary precautions to protect the cities of Greece from infection.

For the account we possess of the doctrines and practice of Soranus we must thank his contemporary Cœlius Aurelianus, who has transmitted to us one of the best medical works descended from the ancients, which contains an account of the accurate descriptions of disease and details of the various methods of treatment then in use. Of all the ancients who have treated that dreadful malady of hydrophobia, none has described the symptoms as fully as Cœlius Aurelianus,[8] who has collected together all he could find on the subject in the works of the Greek physicians.

In his book on chronic disorders, he speaks of the efficacy of music as a remedy in the setting of sciatica.[9] He also mentions its use in epilepsy and adds that the first introduction of this kind of treatment was ascribed to Pythagoras.[10]

Heliodorus, Antyllus, and Leonidas

Considerable improvements were also made in surgery during the reign of Trajan, especially by Heliodorus and Antyllus. The former of these was an eminent surgeon in Rome, several of whose practical observations on diseases of the bones and injuries of the head have been preserved by Nicetas.

Antyllus, a surgeon and contemporary of Heliodorus, was the first writer to describe the method of extracting a cataract. He recommends that a bronchotomy be performed in the setting of sore throat so severe that the accompanying inflammation and swelling threaten the patient with suffocation. He has given detailed directions to guide the young practitioner in selecting veins for bleeding. Leonidas of Alexandria, of whose surgical practice some detailes have been preserved by Aētius, also seems to have been a bold and skillful operator, whose method of proceeding in cases of fistula does not appear to have differed materially from the practice recommended by Pott. He removed cancer of the breast by amputation and cautery.

[8] Cœl. Aurel. de morbis acutis, Lib. iii.

[9] Cœl. Aurel. de morbis chronicus. Lib. v. cap. 1. Quæ eum saltum sumerent palpitando, discusso dolore, mitescerent.

[10] Cœl. Aurel. in loco citato.

Progress in the Field of Anatomy

Anatomy, although neglected in general by the Methodists, was much cultivated by practitioners of the other sects during the first century. Rufus, a native of Ephesus, who flourished towards the close of this century and commencement of the next under the Emperors Nerva and Trajan, is commended by Galen as a skillful physician, and devoted much of his time and attention to anatomical studies. His dissections, however, appear to have been confined to comparative anatomy. He seems to have dissected apes the most often. He successfully traced the course of several of the nerves from their origin in the brain to their distribution throughout the body, and marked the distinction between those of sensation and those of voluntary motion. He exhibited the decussation of the optic nerves at the infundibulum and described the capsule of the crystalline lens. He regarded the heart as the seat of life and the organ of pulsation, and pointed out the difference of structure and capacity between the right and left ventricles. He considered the spleen as having no particular use. Rufus wrote a treatise on the *atra bilis*, or black bile, which the ancients supposed to produce melancholy. However, neither this treatise nor several other works of his quoted by Suidas have reached us.

Anatomy, however, is said to have been still more indebted to the labors of Marinus, whom Galen considered the restorer of the study. Marinus was the author of numerous works, such as a treatise on the anatomy of the muscles, but the whole of these have been lost. Galen is believed to have profited considerably by reading them. Marinus paid great attention to the absorbent system and discovered in the course of his investigations the glands of the mesentery. He divided the nerves into seven pairs and gave the first description of the fourth pair or *nervus palatinus*. He is also said to have discovered the *par vagum*, which he denominated the sixth pair. Caius Plinius Secundus, the illustrious natural historian of Italy who was born at Verona and flourished in the time of Vespasian, though not strictly an anatomical writer, merits notice among the writers of that class on account of the many curious observations on both human and comparative anatomy scattered throughout his works. However, from not being accustomed to dissection himself and from giving too ready a belief to the statements of others, truth and falsehood are frequently found blended together in his works.

Progress in the Field of Botany

Botany, though hardly aspiring at this period to the dignity of a science, was not altogether neglected. A treatise on this subject and the *materia medica* written by Dioscorides of Anazarba obtained deserved celebrity. Experience has confirmed the justice of his remarks upon the efficacy of several valuable remedies furnished by the vegetable kingdom, such as the decoction of elm bark for cutaneous diseases and polypodium for worm infections. His descriptions of some of the more important drugs merit the highest praise for their accuracy and fidelity. While the other branches of medical knowledge were advancing at a slower yet steady pace, the equally important branches of the *materia medica* and pharmacy were far from being overlooked, and many of the preparations that exist presently in our pharmacopoeias trace their origin to this period. Among the medical writers of the first century, none ranked more deservedly high than Aurelius Cornelius Celsus,[11] who has been called by some the "Latin Hippocrates".

Celsus

From the nature of his writings, Celsus (Figure 10) is believed to have practiced physic, though some have expressed their doubts about this conclusion. From evident acquaintance, however, with professional subjects and, above all, from his work on the human skeleton, in which he describes with equal accuracy and minuteness the form, position, and connection of every bone in the body, we may conclude that he practiced some of the branches of medicine. It appears that he occasionally occupied himself with dissections. His medical writings are regarded as inferior only to those of Hippocrates, over which they possess the advantage that they have descended to us free from interpolations and corruptions of the text. Celsus has freely availed himself of the labors of his illustrious predecessor, yet he did not scruple to profit by the information contained in the writings of others and even joined Asclepiades in laughing at the doctrine of critical days in disease, which he regarded as a relic of the superstitious doctrine of numbers introduced by Pythagoras from the East. He also disagreed with Hippocrates on the subject of bleeding, which he employed more generally but at the same time, more sparingly during each operation, fearing to exhaust the patient too much. He also employed cupping and scarifying more often.

Like the Methodists, Celsus prescribes abstinence during the early stages of disease, leaving the determination of its length and severity to be governed by the circumstances of age and constitution, severity of the attack, nature of the climate, season of the year, and other factors. In the later stages, he advised the patient to partake of a generous, but cautiously regulated, diet. He regarded the pulse as a fallacious and uncertain indication of the state of health or disease because its frequency or slowness varied much with the age, sex, and constitution of the patient. Celsus was not altogether free from some of the popular prejudices of his day, especially with respect to the supposed origin of disease, which he refers directly to the anger of the gods. Regarding the poison of serpents, he appears to have been aware of the innocuous character of the virus when not mixed with the blood, for he recommends extracting it from the wound by sucking, provided that the person who performs the operation has no wound or ulcer in his mouth. Celsus seems to have been the first who recommended immersion in water as a cure for hydrophobia. Celsus appears to have regarded the disorders of the rich as more difficult of cure in general than those of the poor, on account of their being less amenable to the directions of the physician. He also preferred keeping the bowels cleared by diet rather than by purgative medicines. He pointed out the propriety of attending to the secretion of urine and comparing the quantity voided in a given time with the quantity of

---

[11] There were three persons of the name of Celsus distinguished for their learning in the annals of Rome, Celsus the Physician, who is supposed to have been born at Rome and to have flourished under the Emperors Tiberius, Caligula, Claudius, and Nero, or between the thirteenth and fifty-fifth years of the first century; Celsus the Lawyer, who was the contemporary of Plutarch and Florus about the middle of the second century or in the time of Adrian and Antoninus Pius; Celsus the Philosopher, who was contemporary with Julius Solinus and Plotinus towards the end of the same century. Horace in his Epistles commemorates a fourth Celsus, who preceded all these in the days of Augustus and may be aptly distinguished by the tile of Celsus the Plagiarist, whom Horace mentions in his Epistle to Florus.

liquid taken in the same space of time, and observed that if the excess be of the urine, a happy termination of the complaint of edema may be hoped for. The works of Celsus exhibit a compendium of the practice both of his own and the preceding ages rather than the result of his personal experience and observation alone. His works touch upon every branch of therapeutics, whether manual, dietetic, or pharmaceutical. Whoever wishes to learn the practice of the ancients before the days of Galen should turn to the fascinating pages of Celsus.

Figure 10. Aulus Aurelius Cornelius Celsus (Image courtesy of Reynolds Historical Library, the University of Alabama at Birmingham).

The Pneumatic and Eclectic Sects

Two other sects that trace their origin to a time right before the organization of the school of Methodists by Themison must be noted. The first is the Pneumatic sect. It was founded by Aretæus, a native of Cilicia, who practiced as a physician in Rome about the middle of the first century. He adopted the system of the Stoics and became an opponent of the doctrines and practice of Asclepiades, Themison, and their followers. He extended the theory of preexistent germs and accounted for the fluctuations of the pulse by referring them to the exhalation of the spirit from the heart and arteries, from which circumstance the sect derived its name. Aretæus paid considerable attention to many branches of pathology, but his reputation arose chiefly from his inquiries on diet and his investigation of atmospheric influence.

The Eclectic, or Episynthetic, sect originated in an attempt made by Agathinus, a pupil of Aretæus, to reconcile the doctrines of his preceptor with those taught by the Empirics and the Methodists. Although he thus virtually became the founder of a new sect, the merit of consolidating its doctrines belongs rather to Archigenes and Aretæus, who took the leading doctrines of the Pneumatic sect and reduced the theory of that sect's founder to a more scientific form and enriched it by a number of valuable observations. Archigenes directed his efforts to reform the language of medicine, but with little success. Many of his practical observations, however, have been preserved by Galen and possess merit. Too great an attachment to subtleties was his great fault. His talents, nevertheless, were respectable.

Aretæus

The exact time at which the celebrated Aretæus lived is uncertain. It seems probable, however, that he wrote in the interval between the commencement of Nero's and that of Domitian's reign, or between the years 54 and 81. So fully did he believe in the indispensable importance of anatomy, not only for the just comprehension of the true causes of disease but also for the rational indications of cure, that he begins almost every chapter with remarks on the structure of the part whose affections form its subject. Thus, he has treated the subject of anatomy more copiously than any of the ancient physicians. In common with Hippocrates and Chrysippus the Stoic, he regarded the heart as the principle of life and strength. He also maintained that the heart was the exciting cause or principle of respiration, being seated in the center of the lungs, which it inspires with a desire for fresh air.

He imagined with Erasistratus that the liver was the source of the veins and little else than a concretion of blood. He conceived the stomach to preside over pleasure and uneasiness and regarded it as the common source of all the faculties because of its proximity to the heart, contributing greatly to the composure or dejection of the mind through its consent with the soul. In common with Erasistratus, he maintained that the nerves were not only the organs of sensation, but also the source of all the action and motion of the limbs. Tetanus he believed was a disease of the nervous system, which he also supposed to be the seat of disease both in gout and encephalitis.

Aretæus applied anatomy, like the Dogmatists, to the explanation of the phenomena of disease and the rational system for its treatment. Epilepsy, as Aretæus informs us,[12] was attributed by many of the ancients to lunar influence entirely, it being supposed that the deity of that planet inflicted it as a punishment upon wicked people for their crimes. He recommends, in cases of inveterate headache, opening the two arteries behind the ears to afford a greater and speedier relief to the head than that afforded by opening the temporal arteries. He has left us an accurate description of a tonsillar abscess,[13] upon which a learned

[12] De diuturnis morbis. Lib. i. cap. 4.
[13] De causis et signis morborum acutorum. Lib. i. cap. 9.

comment has been written by Severinus. He is also the first medical writer who has remarked on the influence that the mind exerts over the health of the body, and the reaction of the body upon the mind. Besides Aretæus, medical science is indebted to Cassius the Iatrosophist, another member of the Eclectic sect, who has written many excellent remarks on the diseases of the nervous system. While the science of medicine was thus advancing at a rapid yet steady pace along the path of improvement, the illustrious Galen arose.

Chapter 3

Galen and His Successors

The Early Genius of Galen

Galen (Figure 11) was the son of Nicon the Geometrician. He was born on an undetermined date in the Asian city of Pergamus, which was celebrated as the place where parchment was first manufactured. He is generally believed to have flourished under the reigns of Trajan[1] and the three succeeding emperors and to have lived to be at least 140-years-old. Having at a young age mastered the logic and philosophy usually taught in his time, he studied medicine under Satyro and Pelops. Galen first appears to have distinguished himself as a medical writer under the reign of Adrian around the year 130, at which time he was noticed as both a teacher and a practitioner. Rivalry prevailed among the Dogmatists, Empirics, Methodists, and other medical sects discussed in the preceding chapter, each of which was supported by numerous and zealous partisans. As a result of various schisms among the Dogmatists, some of whom followed Hippocrates and others Erasistratus or Asclepiades, the Dogmatists' strength was greatly weakened, and the Methodists came to be held with highest esteem from the public, whereas the Empirics fell into contempt, which their idle pretensions, inordinate vanity, and irrational practice merited.

Such was the state of medical science and its affiliated parties when Galen appeared on the scene to purge the art of healing of many errors that had been incorporated into it since the days of Hippocrates. Galen is thought to have joined the Eclectics, but to which of the prevailing sects he originally claimed membership is not certain. The Eclectics were so named for their disclaiming exclusive preference for one writer over another and professing to select from every author indiscriminately whatever appeared to be most valuable. Galen commenced his medical career with a similar profession of impartiality, thereby making his membership to the Eclectics more probable. However, he eventually attached himself almost exclusively to the doctrines of Hippocrates and thus strayed far from his initial commitment to impartiality. His devotion was suggestive of strong bias, such that it frequently led him into the greatest injustice to the merits of others. Although he surpassed both Hippocrates and Aretæus in knowledge, he fell considerably below the level of either with respect to

[1] Trajan succeeded Nerva in the year 98, to which if we add 140, the number of years Galen is reputed to have lived, it will bring us down to the year 238, or 77 years later than the death of Antoninus Pius, the latest of the three Emperors under whom he is said to have flourished.

observational skill. This shortcoming resulted not from any deficiency in his intellectual prowess, but from his adopting erroneous theories.

Figure 11. Claudius Galen, 130-200 A.D. (Image courtesy of Reynolds Historical Library, the University of Alabama at Birmingham).

Galen's Admiration of Hippocrates

Galen began to detect the futility of the prevailing systems at a young age. He was dissatisfied with what his masters taught him as incontrovertible truths and immutable principles, and he was filled as it were with a new light upon studying the works of Hippocrates, which he came to admire as he did the operations of Nature. Thus, he came to rely exclusively on the guidance of Nature and Hippocrates, and took upon himself the labor of commenting upon the writings and expounding the doctrines of Hippocrates. Thus, he revived the Hippocratic system of medicine, which had been debased by the Dogmatists. Galen does not wholly escape the charge of contributing to the adulteration of the genuine text, as Cabanis declares that, although Galen contributed to the Hippocratic system a luster that it did not originally possess, "what it gained in his [Galen's] hands must be allowed to have more the appearance of dress and ornament than of real solid acquisition. The observations which had been collected, and the rules which had been laid down, by Hippocrates, in assuming a more splendid and systematic form, lost much of their original purity; Nature, whom the Coan Physician had always followed with so much accuracy and caution, became obscured, and as it were overwhelmed, beneath the foreign pomp of various sciences and dogmas; and the art of medicine, overloaded as it already was with subtle and superfluous rules, became entangled in new and needless difficulties."[2]

Galen, while professing to restore the doctrines of Hippocrates to their original purity, in reality revived the opinions of the Dogmatists. Hippocrates founded his doctrines upon observation, whereas the Dogmatists spurned observation and substituted for observation their own unsubstantiated theories. Therefore, although avowedly taking the opinions of Hippocrates for his guide, Galen founded his own system, which was influenced more by the pseudo-Hippocratic doctrines of the Dogmatists than by the authentic doctrines of Hippocrates. Notwithstanding the numerous commentators who had preceded him in the task of illustrating the writings of Hippocrates, Galen denied that any but he himself had been able to comprehend their meaning.

Galen's Definition of Medicine

"Medicine," according to Galen, "is a science which teaches what is sound and what is not so; and what is of an indifferent nature, or holds a middle place between what is sound and what is the reverse." In one of his books, however, Galen defines medicine to be an art that teaches the preservation of health and the cure of disease. He says that three things constitute the object of medicine, and that the physician ought to consider them as sound, unsound, or neutral or indifferent, the body itself, signs, and causes. He regards the body to be sound when it is in a good state with respect to the simple parts of which it consists and when there is a just proportion among the organs formed from these simple parts. He considers the body to be unsound when it departs from this state and from the just proportion of its organs, and to be indifferent or neutral when it is in a state between these two extremes. Signs that indicate both present health and the prospect of its continuance he terms salutary, and signs that denote either ill health or its approach he terms insalubrious. Signs that are not indicative of health, disease, or the prospect of either he terms neutral or indifferent. He classifies causes in the same manner.

Galen contends that these three states encompass all the gradations between the two extremes of health and sickness. Each state has its prescribed limits. From this definition, a perfectly sound state of the body appears to be rare, but we are not the less able to conceptualize such a model so as to guide our judgment with respect to different constitutions. Based on this principle, Galen establishes eight other leading constitutions, each of which varies more or less from the assumed type of a perfect one.

Besides these constitutions, he imagined several others to exist that arise from hidden causes and proceed from what he terms the idiosyncrasia, or idiosyncrasy of the constitution. It is this idiosyncrasy that causes the particular aversions peculiar to individuals and inexplicable upon any other supposition. However, notwithstanding how these several

[2] Coup d'Œil sur les revolutions et sur la reforme de lit Médecine.

constitutions differ from the perfect model assumed for comparison, this deviation from health should not be regarded as amounting to positive disease, or that all in whom it occurs are to be viewed as individuals who are unduly anxious about their health. The degree of divergence constitutes disease, not divergence itself. Galen derived the signs or indications of a good, bad, or indifferent constitution from the original qualities of hot and cold, moist and dry, and from their proportion or disproportion to the bulk, figure, and situation of the organic parts.

Galen endorsed the Hippocratic doctrine of the four humors, which are blood, phlegm, yellow bile, and black bile. He also acknowledged three kinds of spirits, which are vital, animal, and natural. The spirits served as instruments for as many sorts of faculties, which resided in the organs where they were formed. The natural, vital, and animal faculties were seated in the liver, heart, and brain, respectively. The natural faculty presided over generation, growth, and nutrition. The vital faculty dispensed warmth and life from the heart to every part of the body through the arteries. The animal faculty, seated in the brain in conjunction with the governing or reasoning faculty, presided over all the other bodily functions and communicated the power of motion and sensation to every part of the body through the nerves. Both Hippocrates and Galen traced the original of all these to the first great cause and called it "nature".

Galen's Pathological Doctrines

These fundamental principles having been laid down, he proceeded to define disease as consisting in "such a preternatural disposition or affection of the parts of the body, as primarily, and of itself, impedes their natural and proper action." Disorders he arranged under three principal classes, the first comprising those of the similar, and the second, those of the organic parts, while the third embraced those common to both. He also followed Hippocrates in his division into acute and chronic. The causes of disease he divided into external and internal, the first of which are six in number, and when properly used contribute to the preservation, otherwise to the derangement, of health. These causes are the passions, air, meat and drink, motion and rest, sleeping and leaking, and retention and excretion. These he named the procatarctic or remote causes because they call into action the internal causes, which are antecedent and conjunct. The first of these consists chiefly in a corrupted state of the humors, arising either from fullness or a viscous state of the humors. Too great a fullness constitutes plethora, which may either affect all the humors of the body together, or only one particular humor that predominates over the rest. Therefore, a plethora may be sanguineous, phlegmatic, bilious, or melancholic. The first differs in that the blood, which is the basis of the other three, may far surpass the rest and constitute a simple plethora, while an excess of any of the others is termed a cacochymy or a redundance of vitiated humors, not plethora, because these humors, abounding in an undue proportion, serve to vitiate and corrupt the blood. He also classified causes of disease as manifest and evident, and latent and obscure.

From the consideration of the causes he proceeds to the symptoms of disease, which he defines to be "a preternatural affection depending upon a disease, or following it as a shadow does a body." He recognizes three kinds of symptoms, the first and most numerous of which consist in an injury, or interruption of the action of the parts; the second in a change of quality, unaccompanied by any disturbance of that action; and the last in defects of excretion and retention.

The signs of disease he arranges under diagnostic and prognostic, the former being employed to point out the distinctions among the several kinds of disease, and the latter those that are collected from the species, symptom severity, and peculiar character of each and that enable us to form some opinion as to its probable duration, effects, and termination. The diagnostic symptoms he divides into pathognomonic or unvarying symptoms, which are characteristic of certain complaints and thereby serve as infallible marks of distinction, and adjunct, which are common to many complaints.

Galen's Achievements

It is sufficiently evident that Galen's system was more theoretical than practical. With all his faults, however, Galen was illustrious, and the services he performed for science will be forever duly appreciated. There is no department of the medical profession which does not exhibit marks of his industry. He is the first writer who appears to have turned his attention to the art of obtaining the aroma of plants and flowers by distillation, or at least he is the first writer who has left behind him a description of the process. He acquaints us[3] with the curious fact that starlings are fattened upon hemlock, which is poisonous to man, and he appears to have been aware of the innocuous character of animal poisons when not mingled with the blood. He mentions the powerful influence that habit exerts over the system in rendering it insensible to the effects of many virulent poisons, as exemplified in the case of a woman who had gradually accustomed herself to the use of hemlock, of which she was able to take considerable quantities without the slightest injury. Diabetes appears to have been a rare disease among the ancients, since we find little notice taken of it by any of the writers who preceded Galen. Even Galen himself acknowledges having seen only two cases of it.[4]

Galen's Contributions to Anatomy and Physiology

Galen contributed to anatomy and physiology by noting facts collected during his various travels and making discoveries that his attention to comparative anatomy enabled him to make. He appears, however, to have been aware of the insufficiency of comparative anatomy to convey an adequate idea of the structure and arrangement of the human viscera. During the period in which he lived, and for a considerable time before, human subjects were not difficult to obtain. Human dissection does not appear to have been neglected by the practitioners and teachers of the age, although it may have been less frequent than in our days. Practitioners in Rome were permitted to dissect the bodies of enemies, criminals, and exposed children, which were all easily obtained. There was no law either to prevent a master from dissecting the body of a deceased slave or selling it to another for the purpose of dissection. We can readily collect from the remarks of Galen himself that he regarded comparative anatomy merely as an introduction to human dissection, without which the latter could not be

[3] Simpl. Medic. Facult. Lib. iii. cap. 18.
[4] De Loc. Affect. Lib. vi. cap. 3.

studied with sufficient advantage. "It is easier," he observes, "for an experienced anatomist to detect at a glance what is familiar to his observation, and what is not, than for a novice to discover by close application what is most obvious." Among other methods of acquiring anatomical knowledge that Galen recommends were those fortuitous opportunities that accident sometimes presents. This "fortuitous anatomy" was the only kind tolerated by the Empirics. For example, he describes an ailment presenting with carbuncles that had been epidemic in Asia and which often afforded the opportunity of observing the underlying muscles where they had been exposed through the effects of the condition. He composed several works on anatomy.

It might have been expected that the labors of Galen would produce an immediate and permanent improvement in the science of medicine, but this unfortunately was not the case. Not long after his death, medical knowledge subsided almost below its former level, and darkness enveloped the globe, or at least the fairest portion of it, for thirteen centuries. However, during this period some few but brilliant glimpses of genius burst upon the gloom and demand our notice.

Oribasius

Pergamus, which boasts of having been the birthplace of the illustrious Galen, also claims the merit of having given birth to Oribasius,[5] who appears to have been born about the year 351 in the time of Julian the Apostate and to have received the first rudiments of his education along with Magnus and Ionicus at the school of Zeno the Cyprian[6] at Sardis. Oribasius is allowed by Eunapius, who is admitted to have possessed considerable skill in medicine, not only to have been the most learned man and most skillful physician of his day, but also to have been pre-eminent for his eloquence. Under Jovian, Julian's successor, he was stripped of all his wealth and splendor, banished from Constantinople, and fell into the hands of the barbarians, in whose regard and veneration he made such rapid progress through his medical skill and amiable qualities, that he was almost worshipped as a god. Being recalled from his banishment, he lived again in affluence and splendor to the period at which Eunapius wrote, which Freind conjectured to have been about the year 400. Both Oribasius and Aetius, who lived about 150 years after him, have preserved in their works many valuable fragments of the ancient writers, such as Archigenes and Herodotus of the Pneumatic sect, and Posidonius and Antyllus. Posidonius was highly commended by Galen. Antyllus wrote several works in which he discussed gymnastics.

Among the fragments of Antyllus preserved by Oribasius, we find many varieties of exercises unnoticed in the works of Galen. Oribasius speaks much of the advantage of scarification,[7] a method of cure rarely noticed by former writers. He mentions having repeatedly found it to be useful in cases of amenorrhea, dyspnea, cephalalgia, and defluxion from the eyes. To Oribasius we are indebted for the first description of that singular kind of insanity known as lycanthropy, in which a person believes that he is an animal. Among the works of Oribasius (Figure 12), although chiefly compilations, we encounter many new facts

[5] Eunapius in Oribasio.
[6] Zeno, as we learn from Julian, Ep. 47. moved to Alexandria where he became a distinguished professor.
[7] Oribasii Coll. 7. 20.

and observations on the subject of diseases. For epilepsy, he prescribes bleeding at the commencement, purging four or five days afterwards, and cupping and scarifying three days later.

Figure 12. Title page from Oribasius Sardianus' *Oribasii medici clarissimi in aphorismos Hippocratis commentaria hactenus non visa*. Patavii, Typis Mattaei Cadorini Superiorum Consensu, 1658.

Aetius

Aetius flourished about two centuries after Oribasius or, as some authorities say, in the beginning of the sixth century. He was born at Amida in Mesopotamia on the banks of the Tigris River. Aetius studied at Alexandria, a circumstance that sufficiently accounts for the number of remarks on the pharmacy of the Egyptians to be found in his works. He made a large collection of formulae, chiefly of potions sold as secrets by their inventors, some of which appear to have been introduced merely for the sake of ridicule. Like his predecessor Oribasius, he preserved many fragments of ancient writers that otherwise would have been lost, especially fragments from the works of Soranus the Methodist and Leonides the Eclectic. Aetius is among the number of those writers who have left us the result of their reading or experience in the treatment of that affliction which follows the bite of a rabid animal. He also gives an account of lycanthropy.

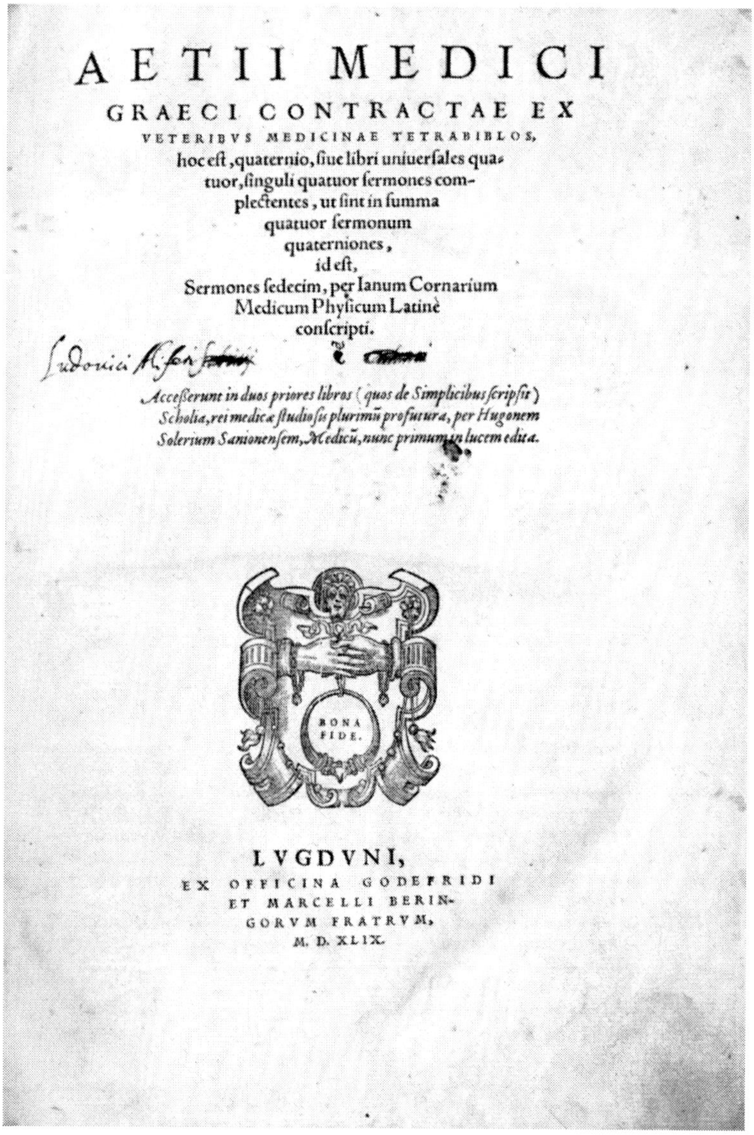

AETII MEDICI
GRAECI CONTRACTAE EX
VETERIBVS MEDICINAE TETRABIBLOS,
hoc eſt, quaternio, ſiue libri uniuerſales quatuor, ſinguli quatuor ſermones complectentes, ut ſint in ſumma quatuor ſermonum quaterniones, id eſt, Sermones ſedecim, per Ianum Cornarium Medicum Phyſicum Latinè conſcripti.

Acceſſerunt in duos priores libros (quos de Simplicibus ſcripſit) Scholia, rei medicæ ſtudioſis plurimũ profutura, per Hugonem Solerium Sanionenſem, Medicũ, nunc primum in lucem edita.

BONA FIDE.

LVGDVNI,
EX OFFICINA GODEFRIDI ET MARCELLI BERINGORVM FRATRVM,
M. D. XLIX.

Figure 13. Title page of Aetius of Amida's book, *Aetij medici graeci Contractae ex veteribus medicinae tetrabiblos*. Lugduni, Ex officina Godefridi et Marcelli Beringorum fratrum, MDXLIX (1549).

Aetius appears from his writings to have been a practical and experienced surgeon, and his works abound in valuable observations on this subject, as well as descriptions of the various methods of performing several operations. He is more copious even than Celsus on the history and treatment of disorders of the eyes. However, he omits the consideration of fractures and luxations, to the discussion of which Celsus has devoted an entire book. Aetius mentions many facts not to be found in the works of Galen or Celsus. The number of manual operations he describes is more than double that found in the works of either of these writers, and he even notices some that are omitted in the works of Paulus. We learn from many parts of his works how frequent the use of actual and potential cautery[8] was among the ancients. Aetius made observations in a chapter principally taken from Herodotus in which, speaking of eruptive diseases of the skin that either accompany or supervene upon a fever, especially when they prove troublesome from itching or effloresce like flea-bites, he says that the system is in such cases loaded with vicious and redundant humors that should be carried off either by emetics or purgatives in order to prevent the vital parts from being injured. At the commencement, when the fever is high, he recommends bleeding. Aetius is the first who gives an account, taken from Leonides, of the Guinea worms. The works of Aetius (Figure 13) abound in external remedies, and he devotes nearly an entire book to the preparation of plasters. He has not only collected all the prescriptions to be found in Galen's commentaries on the preparation of medicines, but also all that could be gleaned from more modern works, whether Greek, Persian, or Egyptian. He appears to have been the first Christian writer among the Greeks who introduced the doctrine of spells, relics, and incantations into medical practice, although such remedies had been much in repute among the ancient Egyptians.

Palladius the Sophist

Palladius the Sophist, or Iatrosophist, received his education at Alexandria. Neither the time nor the place of his birth is distinctly known. He was evidently junior to Aetius, if not to Alexander, who flourished many centuries later. He was the author of some commentaries on the book of fractures written by Hippocrates. Of his commentaries on the sixth book of epidemics, only seven sections have been preserved. In these commentaries, he explains both Hippocrates and Galen with considerable clearness and precision, and particularly remarks upon the increased prevalence of calculus in his time and the increased difficulty of cure, which he ascribes to the growth of luxury and the greater prevalence of indolent habits. To note, little confidence should be placed in the title pages that we frequently find prefixed to ancient manuscripts, for the name of the author is often inserted at random from mere conjecture, erroneous information, or some private suggestion of the librarian without any regard to truth.

Alexander of Trallis

Alexander was born at Trallis in Lydia. Alexander flourished in the time of Justinian, between the years 527 and 565. He was a writer of far more originality than either Aetius or Oribasius. He was the son of Stephanus, a physician of that city. He was instructed under his father Stephanus and under the father of Cosmas in the first rudiments of medicine. It is uncertain whether he was a Christian or a Jew, or to which medical sect he belonged. Fabricius founds an idle conjecture as to his having been a Methodist. His writings, however, show him to have been directly opposed to the tenets of that sect. Furthermore, Alexander

[8] Actual cautery was performed by means of a heated iron and was much quicker in producing its effect than potential cautery, which consisted of a caustic application of pure potash combined with quick-lime.

never once names the grand division of diseases established by the Methodists, and his constant use of purgatives in almost every complaint goes against the precepts and practice of the Methodists. In addition, the distinction of sects cannot be traced to before the days of Galen, not even to those of Alexandria. Galen established the Dogmatic sect so firmly that all the others soon became extinct, or rather were absorbed into it. The system founded by Galen could not, however, be termed a sect, as it was based upon all that was valuable and carefully selected from the other sects, including the Dogmatic sect. Thus, the modern Dogmatists and the Methodists agreed closely upon many points, especially regarding the treatment of diseases, thereby making it difficult to distinguish between them.

While other writers adopted no method for arranging diseases, he classified them with the most symmetric order, from the beginning to the end. Hence, he is the only Greek writer who can enter the lists in this respect with Aretæus. It is bizarre that Alexander takes no notice whatsoever of the diseases of females. However, as he treated, like Aretæus, a small number of complaints, we may reasonably conclude that his practice did not extend to those particular complaints, which form in general a distinct branch of the medical profession.

Alexander is admitted by all to have been a physician of the highest skill, celebrated not only in Rome, but also in every country to which he traveled. He was referred to as Alexander the Physician for his excellence. In the therapeutic parts of his works, he is clearer than his predecessors from the aspect of his formulating his own medicines and being in consequence better able to discover their effects from repeated observations, as he explains in the preface to his twelfth book. He placed great faith in the powers of all his own medicines and had a superstitious reliance upon the efficacy of charms and incantations. With this predisposition to superstition, it is not surprising that Alexander should have applied himself to the study of magic. He furnishes perhaps the only instance of a physician who borrowed anything from Osthanes, one of the most ancient of the Persian magi. In other respects, although an admirer of the ancients, he did not hesitate to express his disagreements with them, sometimes even questioning the opinions of Galen.

He excelled at diagnostics, especially in pointing out the distinctions between those complaints that most nearly resemble each other. Alexander appears to have written a work on disorders of the eyes and another on fractures, both of which have been lost. With this exception, he has confined himself wholly to describing the history, symptoms, and treatment of internal complaints without diverging, like most of his predecessors and contemporaries, to the subject of anatomy, surgery, or pharmacy. He has given an entire book on the history of gout, a complaint wholly unnoticed by Galen. He describes inflammation of the brain with the greatest accuracy, and accurately refutes the preposterous opinions of those who ascribed it to a problem of the diaphragm. His description of the various kinds of melancholy is remarkable for its clearness and precision. For mumps, he advises the use of the scalpel before the exhibition of any other remedies. Alexander takes notice of tubercles in the lungs accompanied with dyspnea but unattended with fever or expectoration. He is perhaps the first writer who has spoken of a stone being coughed up from the lungs, producing all the symptoms of true pulmonary consumption by the irritation it occasioned. Bulimia, or inordinate appetite, he observes to be sometimes the result of worms and mentions the case of a woman afflicted with worms who complained of a perpetual gnawing in the stomach and headache. Alexander has left 12 books, which are extant and some of which he wrote at the request of Cosmas out of gratitude to his father.

Jacobus Polychrestus, Uranius, and Procopius

Alexander mentions several contemporaries of his own profession who appear to have been men of talent and eminence. He particularly praises Jacobus Polychrestus for his singular excellence both in physic and philosophy. Polychrestus received his education from his father Hesychius, whose zeal in the pursuit of knowledge led him to visit many countries. He was a native of Alexandria and was appointed first physician to the emperor Leo the Great. Polychrestus frequently employed enemas and suppositories. He was sparing in the use of cautery and the knife in surgical cases, and was by no means an advocate for bleeding.

Another of Alexander's contemporaries, Uranius, also was remarkable in his character. He was a native of Syria and practiced physic at Constantinople. He appears to have had, like Thessalus, more assurance than learning and more conceit than skill. He chiefly frequented the booksellers' shops and the public piazzas near the court, where he used to collect a number of persons of his own stamp, equally deficient in knowledge, talents, and morality. Such were the persons with whom he was daily in the habit of disputing the most sacred and important subjects, such as the essence and attributes of the deity, the formation of the world, and other matters equally above his comprehension and beyond his knowledge. Uranius's ignorance betrayed him into a thousand inconsistencies and involved him in numerous dilemmas. When Arebindus was appointed ambassador to the court of Persia, Uranius found means to get himself included in the Ambassador's train, and so far succeeded in concealing his natural defects and assuming the demeanor of a sage that, on his first appearance before the Persian monarch, he made so favorable an impression that Chosroes assembled all the magi to dispute with him. Despite the abstruse nature of the questions discussed and his utter ignorance of them, Uranius succeeded through his prodigious effrontery in silencing them. Even after quitting the court of Persia, Uranius was very fortunate to retain the friendship of Chosroes, who even honored him with his correspondence, a circumstance which raised Uranius's vanity to the most extravagant pitch and caused him to treat all who were beneath the rank of princes with contempt.

Procopius was a celebrated historian and a native of Palestine who wrote in the time of Justinian. He was also a contemporary of Alexander and was imagined to be a physician. We are not in possession of any evidence in support of this, with the exception of passages in some of his works in which he has treated some medical subjects with a minuteness and accuracy of detail suggesting a practical acquaintance with them. He embraces every opportunity to speak favorably of the professors of medicine. In noticing some surgical matters, Procopius speaks like a person who had a practical acquaintance with the subject of which he treats and not like a mere historian who records the statements of others. When describing the wound that proved fatal to Artabazes, he comments that an artery in the neck was divided and caused a hemorrhage beyond their power to control.[9] He gives us a full account of the wound Arses received in the face and elaborates upon the perplexity of the surgeons, who were at a loss as to how to extract the arrow without the loss of the eye and serious injury to the nerves and membranes. A notable historical account by Procopius is his description of the plague that devastated Constantinople in the year 543. This work could hardly have been written by a person who had not studied medicine.

[9] Bell. Goth. Lib. iii.

Paulus

Paulus (Figure 14) was born on the island of Aegineta , sometime during the seventh century. We are unclear as to where he received his education, but may reasonably conclude it to have been Alexandria, which continued to flourish as a school of medicine for a considerable period after its capture by the Arabs. Paulus, after completing the first rudiments of his education, travelled to many countries in quest of improvement. He wrote upon the diseases unique to females and is the first who has given us a treatise on midwifery. Thus, he has been called Paulus Obstetricius, or Paulus the Accoucheur, by the Arabs. He was a great compiler and has preserved for us many fragments of the ancients that would otherwise have been lost, in particular the letter written by Diocles to Antigonus on the subject of preserving health. He differed materially both from Oribasius and Aetius. He borrowed freely from the writings of Alexander, not only taking the substance but often transcribing the very words.

Figure 14. Paulus Aegineta (Image courtesy of the National Library of Medicine).

Paulus chiefly distinguished himself in the practice of surgery. He was in the habit of performing with success the most difficult and dangerous operations. His sixth book, which recounts the various operations of surgery, is the best treatise on the subject before the revival of letters and contains many valuable original observations. From the manner in which he speaks of these operations, it is obvious that he was practically acquainted with the method of performing them and did not merely transcribe the descriptions of others. He even dissents from the opinions of previous writers, such as Galen, from whom he differs on the subject of aneurysm. He questions the plan recommended by Hippocrates in cases of fracture of the nasal bones and disagrees with Leonides about the treatment of a hernia varicosa. His fourth book details the different external applications required for ulcers, wounds, and other blemishes that do not require a manual operation. He not only gives us a faithful account of the several methods of treatment practiced by the ancients, but also of those that he adopted himself. Paulus has also left a detailed account of a multitude of perfectly new operations, as will appear upon a comparison of his works with those of the Latin Hippocrates. Paulus describes the operation for removing the breasts in men when they grow to an inconvenient size. This operation is not noticed by any other writer.

Stephen of Alexandria

Stephen is believed to have been born either in Athens or Alexandria. He appears, from the various evidences collected, to have flourished around the beginning of the seventh century under the reign of Heraclius. He is conjectured to have been a pupil of Theophilus, a fact respecting which the evidence adduced is by no means satisfactory. If we admit him to have been the same who was sometimes known by the name of Stephen the Chemist, his work *de Chrysopoeia* clearly shows the age in which he wrote in its dedication to the emperor Heraclius. It is, however, probable that there were more than one eminent practitioner of the name, and that the works of several distinct individuals have, in the confusion of the darker ages, been erroneously ascribed to one. Abi Osbeia speaks of seven physicians resident in Alexandria, and among them one named Stephen, who digested the voluminous works of Galen into sixteen books, and these again into seven classes according to the nature of their subjects.

Theophilus

Theophilus (Figure 15) flourished about a century later than Paulus, between the years 610 and 641. He is spoken of by the names of Iatrosophista, Protospatarius, and Monachus. It is to this writer that one of the manuscript copies of the work on fevers written by Palladius and preserved in the Imperial Library at Vienna is erroneously ascribed. Theophilus is the first who has written expressly on the subject of urine and given a rational explanation for the causes of its variations in color and consistency, as well as the prognoses to be drawn from them. He has also written on the subject of intestinal evacuations. A treatise on the pulse, which has been translated under the name of Philaretus, is ascribed to Theophilus. It is, however, a mere epitome of Galen's book on the same subject. Besides these works, Theophilus was the author of five books on the structure of the human frame, of which several editions have been published both in Greek and Latin. This work also is for the most part an epitome of Galen's treatise *De usu partium*, but it contains many original facts and observations. For example, Theophilus is the first to trace the olfactory nerves from their origin in the brain to their extensions to the membrane lining the nostrils, so as to form the organ of smell. He was the first to demonstrate the strong ligament that binds together all the articulations of the vertebra and prevents their dislocation. Theophilus wrote a book of commentaries on the aphorisms of Hippocrates, which displays an equal acquaintance with the opinions of Aristotle and the doctrines of Hippocrates.

Figure 15. Theophilus, from "*De urinarum differencia negocium*" (Image courtesy of the National Library of Medicine).

An opinion has long prevailed, but certainly without sufficient foundation, that no improvement in the science of medicine took place for a period of more than twelve hundred years after the death of Galen. The origin of this erroneous opinion is to be found in the circumstance of the various writers who succeeded that great man, having confined themselves chiefly to transcribing and commenting upon his works to the exclusion of almost every original idea or attempted discovery. However, we have already seen in the course of the present chapter that the works of Oribasius, Aetius, Palladius, Alexander of Trallis, and Paulus contain a vast variety of new and original facts, observations, and arguments. The amalgamation of all the sects into one, which the genius of Galen inspired, had the happy effect of checking that absurd propensity to indulge in visionary speculations that tended to embarrass rather than promote science. As the spirit of theorizing was repressed, the attention of practitioners was directed to the more substantially useful parts of the medical profession, thereby improving medical knowledge. In corroboration of this assertion, we shall find in the works of the authors just mentioned descriptions of diseases unnoticed by their predecessors, important improvements in the practice of treating those diseases already known, accounts of new medicines introduced into practice, and large additions to the operations of surgery.

Chapter 4

The Influence of Muslims and Arabs on Medicine after the Capture of Alexandria

The Origin of Smallpox

After the capture of Alexandria by Muslims and the establishment of the Islamic Caliphate in the Middle East, the subsequent political and socioeconomic alterations significantly influenced the progression and advancement of all branches of knowledge, especially medicine. Early in this period, a great effort was made to collect and translate Greek textbooks into Syriac and Arabic. Several hospitals and colleges, almost entirely in conjunction with mosques, were established throughout the Caliphate, especially in Baghdad.

Before we proceed to speak of the fall of Alexandria or its influences, it is proper to pause a little for the purpose of noticing the first recorded appearance of smallpox. We know that smallpox was reported and known to the Chinese about 1200 BC, but the records of the Chinese partake too much of the apocryphal character of romance to furnish materials for serious argument or credible discussion. The earliest mention of this disease, upon which reliance can be placed, is an ancient Arabic manuscript preserved in the public library at Leyden, in which one can find the following words, "This year [late in the sixth century A.D.], in fine, smallpox and measles made their first appearance in Arabia." The first writer who appears to have given any systematic account of smallpox, with the diagnosis of the several varieties and their methods of treatment, is Aaron, who was a native of Alexandria. He was learned priest and physician who flourished about the year 621.

The Capture of Alexandria

The capture of Alexandria in 641, three years after the invasion of Egypt by General Amrou, who commanded the troops of the Caliph Omar, brought the Arabs who were more intimately acquainted with the books and literature of the Europeans, and thus contributed to the preservation of that holy spark of ancient knowledge that would have been wholly extinguished if not for them. The capture of Alexandria hastened the transfer of the knowledge and books of European philosophers to the shores of Asia.

The capture of Alexandria has been believed to have been followed by the destruction of its literary treasures during the reign of Caliph Omar, who is claimed as having declared, in reply to a dispatch from his general, the library to be destroyed. The validity of this account given by Gregory Abulpharagius, who was born in Malatia in 1226, is questionable, as Abulpharagius was born 586 years after the capture of Alexandria and was far less likely to be well informed on this subject. Even the slightest mention of the destruction of the library of Alexandria is not found in the writings of Eutychius, the patriarch of Alexandria who flourished around the ninth century, or George Elmacin, the historian who flourished in the early twelfth century. Both were Christians and natives of Egypt and born long before Abulpharagius, and were consequently far more likely to be well informed on the subject.

Academia and Scholarship in Alexandria

The fact that a vast multitude of the most valuable works of the ancients survived after the capture of Alexandria deserves notice. The preservation of these works was favored not only through the good sense and liberality of the more enlightened among the conquerors, but also through the perseverance of a multitude of men of letters, among whom John Philoponus, or John the Grammarian, was well known. The preserved manuscripts were transcribed and dispersed in various directions. What must have contributed significantly to the preservation of these treasures at such a time was the strong attachment of the Arab conquerors to medical studies and Muhammad's enthusiasm about the study of medicine, having himself written a book of aphorisms on the subject. However, this work has not survived to the present. We also find that Alexandria retained its celebrity as a school of science, especially medical science, for a considerable period after its conquest by the Muslims. Theodunus and Theodocus, two illustrious physicians and eminent teachers of medicine, flourished as late as the close of the seventh century and appear to have been natives of Alexandria. Some pupils of Theodunus and Theodocus were still living in the eighth century, during which the Abbasids took power from the Caliphate. Abi Osbaia wrote a book on the lives of many physicians and informs us that Elkenanus was originally a Christian and a teacher in the schools of Alexandria, but converted to Islam through the persuasions of the Caliph Abdul Aziz. We learn that literature survived at Alexandria, and even the transfer of the schools to Antioch and Harran led to the spread of this knowledge to other parts of the Arab dominions. Alexandrian physicians were respected by the caliphs, for such was the reputation of the Patriarch of Alexandria for his medical skills as late as the start of the ninth century that he was summoned by Caliph Haroun Al-Rashid to the sick bed of one of his favorite concubines.

Aaron and Masarjawaih

The first translations of writings by Greek writers after the fall of Alexandria appear to have been in the Syriac language. Aaron, a presbyter of Alexandria who flourished around the year 622, was the author of 30 books chiefly compiled from the Greeks and named by him *Pandects*. Aaron died around the year 665 or 666. Haly Abbas criticizes him for being too concise in his remarks on the naturals and non-naturals and silent altogether on surgery. The first Arabic translator of the Greek writers was a Jewish physician by the name of Masarjawaih, a native of Syria. He published an Arabic version of Aaron's *Pandects* around the year 683. The majority of succeeding translators followed this example of translating from the Syriac, rather than directly from the Greek. Shortly after this period, literature in Arabia received a fresh stimulus from the support of the Caliph Al-Mansur, the second Caliph of Abbasids. He was a zealous promoter of every useful science, especially astronomy. This caliph founded the city of Baghdad in the year 767 and made it the seat of the caliphs.

Bachtishua

The Caliph Al-Mansur became dangerously ill and sent for George Bachtishua, an Indian physician and Christian who was deeply versed both in the Persian and Arabic languages. At the caliph's request, Bachtishua made a multitude of Arabic translations from the best medical textbooks. Bachtishua was educated and resided in Jondisabur, a city founded in 272 by Shapur I, King of Persia, in honor of his queen who was a daughter of the Roman emperor Aurelian. In this city, medical science had flourished uninterruptedly from the date of its foundation to the days of Bachtishua. Aurelian had sent a number of Greek physicians to Shapur, by whom the knowledge of medicine was introduced and disseminated throughout the East. Al-Mansur received George Bachtishua most graciously. At this time, as in the days of Hippocrates, the practice of medicine was to a great degree confined to private families. George brought up his son Gabriel in his own profession. Gabriel was so successful in his practice that in three days he cured Giafar, first counselor to the Caliph Al-Rashid, of an illness that had baffled even the skill of his father, and afterwards healed a concubine of Al-Rashid who had been under the care of many other physicians without success. For his efforts, he was appointed chief physician to Al-Rashid. After Gabriel, we find at least three to four generations of the family of Bachtishua with members who attained high professional distinction, and many of them translated a great number of medical works into Syriac and Arabic. The Caliph Al-Mansur, having died at the age of 63 years while on a pilgrimage to Mecca, was succeeded by Al-Mahdi, during whose reign Theophilus of Edessa, a Maronite and a distinguished astronomer, made an elegant translation of Homer's *Iliad* into Syriac.

Honain

Honain was born in Hira, Iraq, about the year 764. He was a Christian, and he translated from Greek a number of works chiefly on the subject of medicine. Having been mistreated by Mesue, he left Baghdad for a time in disgust and travelled to Greece or to countries inhabited by Greeks, where he resided for about two years. He perfected his Greek and collected books on every branch of science. He returned to Baghdad and stayed there briefly, and then moved to Bassorah in Persia where he made himself a master of Arabic. Returning a second time to Baghdad, he acquired a great reputation there for his skill in languages and devoted his time chiefly to the translation of Greek works into Arabic. Among these works were the seven books written by Paulus. There is a story related by Abi Osbaia, an Arab biographer and historian, of the Caliph Al-Ma'mun having been visited in his dreams by an old man who called himself Aristotle. The Caliph demanded to know who Aristotle was, and he was informed that he had been one of the most illustrious philosophers of Greece. He commanded Honain immediately to translate his works into Arabic, promising him an equal weight of gold for every volume of his translation completed. Abi Osbaia devoted an entire chapter of his work to an account of at least forty-six Arabic translators, none of whom approached Honain in reputation. Honain, who must have been born during the caliphate of Al-Mansur, flourished under Al-Mahdi, Haroun Al-Rashid, Al-Amin, and Al-Ma'mun. Abi Osbaia tells us that Honain lived to be 100 years old. His son Isaac and his grandson Hobaish also translated the works of Aristotle, Hippocrates, and Galen.

Caliph Haroun Al-Rashid

In the year 792, Haroun Al-Rashid, the fifth Abbasid caliph, ascended the throne of his ancestors and distinguished himself as a patron of literature and science. He adorned the city of Baghdad with a multitude of mosques, public schools, and hospitals, and thus set an example that was long followed among the Arabs of accompanying the foundation of every mosque with the establishment of a college and a hospital. The Moors most rigidly adhered to this practice in Spain during the period of their reign. Such was the state of improvement of the arts under Haroun Al-Rashid that, we are informed, he presented a striking clock to Emperor Charlemagne, King of the Franks. This clock was the first ever seen in France or probably even Europe. Elmacin relates a story about a dangerous attack of apoplexy that Caliph Al-Rashid was inflicted with in the year 802. When a consultation was made, a young but able physician of the distinguished family of Bachtishua alone had the courage to propose a novel operation of opening a vein as the only effective mode of treatment. This measure was strongly objected to by Al-Amin, Haroun's eldest son, but was successfully advocated by his younger son, Al-Ma'mun, who maintained that the proposed remedy was most undoubtedly deserving of a trial. The advice of Bachtishua prevailed, and the operation was performed with a good outcome. The caliph was instantly relieved by the loss of blood and lived a considerable time thereafter.

Mesue

During the reign of Haroun Al-Rashid, a distinguished medical professor by the name of Mesue flourished in Baghdad. He was employed for more than 40 years by this caliph and his successors in expounding and commenting upon the works of the ancient physicians. Mesue was a Christian of the Nestorian sect and born in Jondisabur, where his father was a pharmacist. He received his medical education under Gabriel, the son of Bachtishua. He is said to have surpassed the whole of his contemporaries in erudition and enjoyed the favor and confidence of every caliph under whom he lived. Abi Osbaia notes 37 works of which Mesue was the author.

Caliph Al-Ma'mun

His eldest son, Al-Amin, who only reigned shortly, succeeded Haroun Al-Rashid. Al-Ma'mun, the second son of Al-Rashid, came to the throne upon the death of his elder brother in the year 840 and thereby became the seventh Abbasid caliph. This generous and enlightened prince far surpassed the whole of his predecessors in restoring and propagating the various branches of knowledge. Before the capture of Alexandria, knowledge of medicine was confined to a small number of practitioners whose acquaintance with science was limited. With the ascension of the family of Abul Abbas al-Saffah to the Caliphate, the Arabs were stimulated academically by the new dynasty. As we have already mentioned, Al-Mansur, the second Abbasid caliph, was the first who promoted science. He paid special attention to natural philosophy and astronomy. Haroun Al-Rashid emulated his predecessors, and by enriching Baghdad with schools and hospitals, contributed largely to academia. However, it was reserved for his son Al-Ma'mun, the seventh Abbasid caliph, to complete what his predecessors had begun. Al-Ma'mun labored with the most indefatigable perseverance to collect the works of the learned from every quarter. Having obtained a vast collection of productions and assembled a conclave of the most eminent translators, he gave them works to translate, stimulating their industry by offering prizes and supporting their endeavors by personally attending their lectures and discussions.

Serapion

Serapion, or Yahya ibn Sarafyun, flourished around the time of Caliph Al-Ma'mun. Haly Abbas places him during the ninth century, a conclusion considerably supported by a reference to the writings of Serapion himself in which we find him speaking of remedies employed by Gabriel, the son of George Bachtishua, Honain, and Mesue and a dentifrice named after the Caliph Al-Ma'mun. Serapion appears to have been born somewhere in the province of Khorassan in Persia and to have been raised in Nishapur. He wrote originally in Syriac, which appears to have been the official language in this region and the language employed by Mesue and other writers of that period. Haly Abbas complains that Serapion does not discuss surgery when writing about the treatment of disease and restricts his discussion of treatment to diet, regimen, and internal remedies, and also criticizes Serapion for omitting various diseases, such as elephantiasis, aneurysm, and varices, in his writing. Serapion frequently borrows from Alexander of Trallis, who appears to have been well known among the Arabs.

Rhazes

Rhazes (Figure 16), one of the most illustrious and prolific of the Arabian writers, flourished towards the close of the ninth and beginning of the tenth centuries. He was born in the year 852 in the Persian city of Rey near the province of Khorassan. He moved to Baghdad at the age of 30 years. He did not study medicine until late in his life. He is extolled by the Arabian writers in the most extravagant terms. Eventually losing his sight, he died in 932 at the age of 80 years. So great was his reputation that out of 100 eminent practitioners in Baghdad at that time, he alone was the chosen physician for the celebrated hospital in that city. He is said to have travelled much in Persia in pursuit of knowledge. He was employed professionally by a number of rulers. He had great relationships with all the botanists, oculists, and surgeons of his day, and bore the reputation of being a profound alchemist.

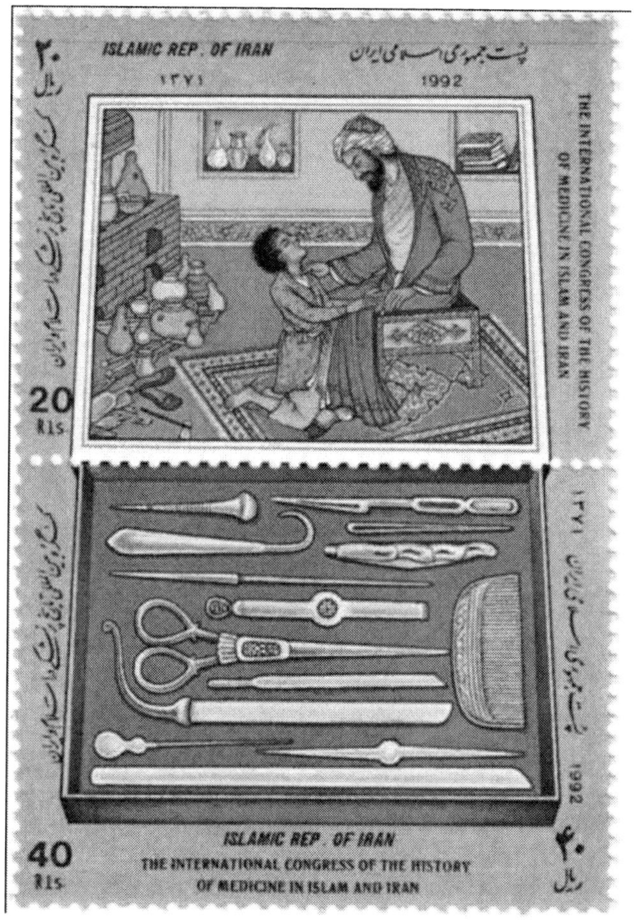

Figure 16. Rhazes as depicted on an Iranian postage stamp.

Abi Osbaia lists at least 226 books of which Rhazes was the author. Rhazes was the author or compiler of the work *Kitab Al-Mansuri*, which he intended as a compendium of medicine in which the materials of another work that was meant for his private use, *Continent*, were more methodically arranged. *Kitab Al-Mansuri* had such a reputation that it was used for ages by most professors as a key textbook during their lectures. Both of these books show a reliance on the Greek writers, leading to some persons unjustly questioning their significance and originality. The accumulation of knowledge is gradual and progressive. It is not the mere accumulation of isolated facts that constitute science, but the digestion of those facts into a connected and intelligible system. Rhazes is the greatest benefactor to the cause of learning. He took up the works of his predecessors in any branch of knowledge, illustrated them by his observations, and enriched them by the incorporation of his own discoveries. Therefore, the excellence of Rhazes as a writer is not to be estimated from a comparison of the matters contained in his works with those in the works of the Greeks, but by the nature of the arrangement and the novelty or importance of the observations that he made. It should also be kept in mind that Greek was not the language of Arabs, and that Rhazes, in rendering the knowledge of the Greeks accessible to his countrymen through the medium of their own language not merely by literal translation but by amalgamation with his own opinions and juxtaposition with his own experience and observations, was a great benefactor to his country. Also, Rhazes is the first medieval writer who has spoken expressly on the diseases of childhood.

Avicenna

Avicenna (Figure 17), the illustrious son of a tax collector, was born at Bokhara in the Persian province of Khorassan around 980. From childhood, he studied mathematics and natural philosophy with so much ardor that, if we believe the statements of his pupil Juzjani, he had gone through the whole of Euclid's *Elements* and a variety of other works on the higher branches of mathematics before the age of 16 years. He then devoted himself with equal zeal to the pursuit of medicine, wherein he made such progress that he soon became as eminent for his skill in medicine as for his proficiency in other branches of science. He resided chiefly in the city of Isfahan and was accused of immoderate indulgence in pleasure, which subjected him to a variety of complaints. As a result of his suspected social transgressions, it became proverbial among his fellow Arabs to say that the study of philosophy did not contribute to virtue and that the study of medicine did not contribute to the preservation of health. Avicenna died in the year 1036 at the age of 56 years and was interred in the city of Hamadan. There is much false speculation regarding the country and age of Avicenna, some of which represent him as a native of Spain and a pupil of Avenzoar, but this account is highly improbable, as Avenzoar himself speaks of Avicenna as his predecessor and praises him for his learning. Others represent him as having been born in Edessa, a city in Syria, in 1145, more than a century after his death.

Avicenna's great work was the *Canon of Medicine*, and such was its reputation throughout the region that it was epitomized and commented upon by a multitude of Arabian writers during the twelfth and thirteenth centuries. Even at an earlier period in Europe, it was so celebrated that no other theories were used in any of the schools of medicine other than those contained in this work, which continued to maintain its ascendency in Europe after the revival of letters. Avicenna was a meticulous observer of clinical medicine. His *Canon* has been written in such a systematic manner that it is stylistically comparable to modern medical textbooks. Each chapter on the diseases of a specific organ begins with anatomy and physiology, followed by a review of the symptoms and signs attributable to the diseases affecting each organ. This discussion is followed by a description of the individual disease with it respective classification, treatment, and prognosis.[1] Although Avicenna has frequently confounded the primary symptoms with those that have no necessary connection with the principal disease, his approach to reporting the manifestations of diseases clearly reflects his attention to detail. He was probably one of the first authors to have adopted a systematic

[1] See, Shoja MM, Tubbs RS, Loukas M, Khalili M, Alakbarli F, Cohen-Gadol AA. Vasovagal syncope in the Canon of Avicenna, the first mention of carotid artery hypersensitivity. Int J Cardiol 2009; 134(3), 297-301.

decision-making approach, which is referred to as evidence-based medicine in the modern day.[2]

Figure 17. Profile of Avicenna reconstructed from the remains of his skull (Image courtesy of Caroline Erolin and Mohammadali Shoja).

Haly Abbas

Haly Abbas was a physician and writer who wrote the fullest and most ancient account of the state of medicine among the Arabs and the most copious history of the Arabian medical writers. He was a man of profound learning for the age in which he lived, and from this reputation obtained the surname of Magus. About the year 980, he wrote a voluminous work designed to be a complete system of medicine (Figure 18), which he called *Al-Maliki* and inscribed with a magnificent dedication to Caliph Adhud al-Dawlah. Stephen of Antioch translated this work into Latin in 1127. The work of Haly Abbas contains many passages that correspond closely with passages quoted by Rhazes as extracts from the work of Isaac, but this coincidence may be accounted for with the supposition of the materials of both having been derived from one common source without affixing to the production of Haly Abbas a

[2] See, Shoja MM, Rashidi MR, Tubbs RS, Etemadi J, Abbasnejad F, Agutter PS. Legacy of Avicenna and evidence-based medicine. Int J Cardiol 2011; 150(3), 243-6.

charge of plagiarizing the works of Isaac. The reason assigned by Haly Abbas for undertaking this work is the inadequacy of all the existing works upon the subject to supply the requisite information, and he even includes Paulus, Oribasius, Galen, and Hippocrates in this sweeping charge of defectiveness.

Figure 18. Title page of the Latin edition of Haly Abbas' book (1523; Image courtesy of Wellcome Library, London).

Abdallatif

Abdallatif, another Arabian physician and historian, whose account of Egypt supplies a valuable link for connecting ancient history with modern history, merits notice for his distinguished talents. He was born in Baghdad around 1161 and studied grammar, rhetoric, history, poetry, and Islamic law in his youth. He commenced his travels around 1189 at the age of 28 years. He visited Damascus and proceeded to Jerusalem, and then went to Egypt, where he formed a friendship with Maimonides. After peace was established between Saladin and the Christians, he waited on that prince at Jerusalem and was received most graciously. While in Jerusalem, he delivered public lectures in the temple. Returning a third time to Damascus, he gave lectures on various subjects, which were well attended. After this engagement, he spent several years travelling through various countries and practicing medicine at different courts. He died in Baghdad, which he had revisited for the purpose of presenting his works to the caliph. He is said to have left behind 150 works on medicine and other subjects.

Medicine among the Arabs in Spain

On the ascension of the Abbasids to the seat of the caliphs in 744, Abd Al-Rahman, the son of Moavia of the exiled Umayyad family, fled to Spain where Arabs had established their dominion some years before. He was received and acknowledged by them as the true and legitimate caliph in opposition to the Abbasid Caliphate, by which he had been expelled from his throne. From the period of this division, an almost chronic animosity commenced between the Arabs of the East, who were centered in Baghdad, and the Arabs of the West, who were centered in Spain, which not only occasioned the latter to forego their dependence upon their motherland, but also the cessation of relations between them. Abd Al-Rahman's family continued to reign in Spain until the period of their expulsion by the Emperor of Morocco around 1030. Abd Al-Rahman founded the city of Cordova and its university, which at one time was the most illustrious in the world and as early as the tenth century boasted to have the largest library in the West. Indeed, such was the rapid progress of learning in Spain under the fostering dominion of the truly enlightened and chivalric Arabian rulers that even in the twelfth century Spain could boast of at least 70 public libraries that were filled with the choicest flowers of literature. Academies were established in Seville, Toledo, and Murcia.

Avenzoar

The family of Avenzoar, or Ibn Zuhr (Figure 19), was ennobled by ancient and splendid descent and by an eminence in the ranks of science. Both his father and grandfather were conspicuous for their success in the practice of medicine. Neither the date nor the place of his birth is known with precision. But from his own works, it is evident that he must have flourished many years after Avicenna, whom he eulogizes in the most extravagant manner. His birth is believed to have taken place towards the end of the eleventh century in the city of Hispal or Seville, the capital of Andalusia. He lived in this city for a long period of his life.

He commenced practice at the early age of 20 years. Avenzoar inherited a handsome fortune from his ancestors, and he practiced gratuitously among the poor and persons of small fortunes, but did not refuse presents from the rich, but he converted these rewards into means of relieving the distressed. He lived to be the benefactor of fellow men, and he toiled to do good and not to accumulate wealth. In addition to being descended from ancestors who had been eminent for their skill as regular practitioners, Avenzoar enjoyed the advantages of a regular and scientific education not confined to the mere practice of medicine, but extended to everything relating to surgery and pharmacy. He wrote a work entitled *Al-Taysīr fil-Mudāwāt wal-Tadbīr*, which contains rules for diet and medicine for almost every disease. We learn from this work to which he later added a supplement titled *Jame* that he was in charge of a hospital and was frequently employed in affairs of importance by the command of Miramamolinus, Emperor of Morocco. Throughout the whole of this work, he uniformly embraces the doctrines of the Dogmatic or Rational sect, which was diametrically opposed to that of the Empirics. Although he treads closely in the footsteps of Galen, his works will be found to lack many principles found in the writings of his predecessors.

Figure 19. Avenzoar (Image courtesy of the National Library of Medicine).

Avenzoar writes about abscess of the pericardium, a disease of which we do not find mention by any other Arab writers or any Greek writers. Salius Diversus, the author of *De febre pestilenti tractatus* and who describes a variety of diseases unnoticed by other writers, devotes an entire chapter to the investigation of this one disease, of which he says he could find no mention made elsewhere. Avenzoar also writes about some other afflictions of the pericardium, such as the thickening of its coats by the production of what he calls a new substance resembling cartilage or pellicles, unnoticed by any previous writer. He also speaks of edema of the pericardium, which, as he observes, Galen does not mention. Avenzoar is the only Arabian writer who speaks of bronchotomy in dangerous cases of tonsillitis. His remarks on the relaxation or obstruction of the esophagus resulting in dysphagia are not to be found in any other Arabian or Greek textbook.

It was not merely in the practice of medicine or surgery that Avenzoar excelled. His skill in pharmacy is testified by his book on pharmaceutical preparations, which discusses both simple and compound medicines and notes valuable novel observations. Avenzoar employed Black Hellebore, or *Helleborus niger*, to treat bony outgrowths from the spine. Being also troubled with a similar painful outgrowth himself, he succeeded in removing it by the use of solvents and purgatives without suffering the slightest inconvenience from its remains. While a boy, Avenzoar took infinite pains to become adept at human anatomy. He not only studied the manner in which the different operations in surgery should be performed, but also performed them with his own hand. Thus, he became equally expert as both an operator and a dissector.

From the apology that Avenzoar has deemed necessary to introduce into his works for his exercising the various branches of medicine, surgery, and pharmacy, it is evident that not only in his time, but also for at least two generations before, these three branches of medicine were practiced independently as three distinct professions. He speaks of several eminent schools of medicine that existed at this period in Spain of which that at Toledo enjoyed the highest reputation. The date of his death is not certain, but according to the most authentic accounts, he died in Morocco around 1217 or 1225. Avenzoar was universally admitted to have been a good and benevolent man. He left a son who followed his profession and rose to an eminence like that of his father. Avenzoar's son was in high favor with Almanzor, Emperor of Morocco.

Averrhoes

Averrhoes (Figure 20) was born in Cordova around 1242. He was a man of sound judgment and powerful talents. He was the son of the high priest and chief judge at Cordova, and was educated at the University of Morocco where he first studied law but then abandoned it for the study of medicine, mathematics, and philosophy. His grandfather appears to have enjoyed considerable distinction among his countrymen. Averrhoes informs us himself in his works that he was acquainted with the sons of Avenzoar, but says nothing about his acquaintance with their father. Having completed his studies, Averrhoes succeeded to the posts of high priest and chief judge after his father's death. Despite the substantial income from these sources as well as that from his professional practice, his generous and liberal disposition kept him perpetually poor. He was soon removed from his prestigious position with charge of heresy and thrown into prison, from which he was released and restored to his hereditary posts in consequence of the misconduct of the person who had been appointed to succeed him. Being accused of impiety, Averrhoes has been represented as denying the immortality of the soul. However, examining the works of Averrhoes himself, one would discover these charges to be utterly lacking a foundation and to have originated solely in the malignancy of his traducers, for Averrhoes has, in one of his treatises, expressly maintained the immateriality of the soul and, in another work, the immortality of the soul.

Figure 20. Averrhoes by P.R. Vigneron (Image courtesy of Wellcome Library, London).

Averrhoes wrote numerous commentaries on the works of Aristotle. At the request of Miramamolinus, Emperor of Morocco, he wrote a work on medicine, which he admits to be only a compilation of the observations of others with some additions of his own. He furnishes nothing new on the subject of anatomy. The practical part of his work is almost equally deficient in novelty. He appears to have been the first to remark that a person who has once had smallpox is not subject to a return of the disease. In the compilation of his great work on medicine, the object that Averrhoes appears to have had principally in view was to rectify the conflicting theories of former writers and to reconcile the conflicting opinions of Aristotle and Galen. He assigned to the former the first and to the latter the second place among the writers and philosophers of the age.

Albucasis

Everything connected with the history of Albucasis, also known as Abu al-Qasim or Al-Zahrawi, is buried in profound obscurity. Not much is known with certainty regarding the time and place of his birth or the place of his education. The slightest notice is not taken of him in any of the Arabian works that have reached us. Some suppose him to have lived to the eleventh century. Finding the practice of surgery to be on the road to demise, Albucasis worked vigorously to restore it to its proper station among the branches of useful knowledge. He criticized those who attempted to practice surgery without first acquiring an adequate proficiency in the knowledge of anatomy and the principles of medicine. His great work entitled *Al-Tasrif* (Figure 21) occupies 32 volumes and is considered by some to display an intimate acquaintance with diagnostics.

LIBER THEORICAE
NECNON PRACTICAE ALSAHARAVII IN PRI-
sco Arabum Medicorum conuentu facile principis: qui vulgo
Açararius dicitur; iam summa diligentia & cura
depromptus in lucem.

Cum priuilegio summi Pontificis
et Imperatoris Romani.

Figure 21. Title page of *Liber theoricae necnon practicae Alsaharavii in prisco arabum medicorum conuentu facile principis* (Latin edition of Kitab al-Tasrif of Albucasis, Abul Qasim, 936-1013.). Cum priuilegio summi Pontificis et Imperatoris Romani, 1519.

This work is chiefly made up of transcripts from Rhazes and other predecessors of his, but it is not destitute of merit. Although he borrows from the Greeks, and especially from Atius and Paulus, he acknowledges only Hippocrates and Galen by name. He is the only writer among the ancients who has left us figures of the various instruments employed in surgery in his day accompanied by a description of the manner in which they were used. He also gives clear and distinct explanations of the nature of each operation and the manner of performing it. He professes to reject everything that is superfluous in surgery and to retain only what is essential, and declares that he will give nothing but the results of his own practice.

The first volume of *Al-Tasrif* is confined to the consideration of cautery. Cautery was more commonly used among the Arabs than among the Greeks, and was performed more commonly by means of caustic potash than by actual burning. The second volume of *Al-Tasrif* describes ninety-six operations performed with the knife, which Albucasis observes to be much more dangerous than cautery due to blood loss. The concluding section of the second volume discusses various methods of venesection and different instruments used for this purpose. Albucasis appears to have been the first who formed a rational plan of treatment for smallpox.

Such was the general progress of medical knowledge among the Arabs during that dark and stormy period of European history. Chemistry and pharmacy were the branches to which they most diligently applied their efforts and to which they introduced the greatest improvements. To them we are indebted for some of the most useful purgatives contained in the *materia medica*, and from them we have borrowed the example of publishing pharmacopoeias.

Chapter 5

Greek Medical Writers from the Tenth to the Fourteenth Centuries

Actuarius

We must now retrace our steps to notice the works of three of the later writers of the Greek school of thought, whose precise age it is difficult to determine satisfactorily and whose works exhibit a style far from the purity of the classical ages. The first is Actuarius, the son of Zachary. His real name was Zacharias, but he appears to have obtained the former name from his position as the chief medical attendant for the imperial court. Authors vary in their placement of him in history; some place him in the eleventh century, others in the thirteenth. Lambecius places him in the fourteenth century. Lambecius contends that Actuarius's work on the art of healing was dedicated to Apocauchus, a notable figure under the Emperors Andronicus the Second[1] and Andronicus the Younger and the Usurper John Cantacuzenus. Apocauchus was a fellow pupil of Actuarius under Joseph Rachendyta, to whom the books *De Spiritibus* are dedicated. Apocauchus was also proficient in philosophy and medicine, if we believe Lambecius, who brings forward evidence from the history of Cantacuzenus, in which Apocauchus is spoken of somewhat ironically. Although the description given by Lambecius may agree on many points with the description of the Apocauchus of whom Actuarius speaks, Lambecius and Actuarius cannot be talking about the same individual, for the Greek writer Nicholaus Myrepsus frequently speaks of and quotes Apocauchus, and Myrepsus lived long before the date assigned by Lambecius to the time of Actuarius.

Myrepsus authored the work *Antidotarium*, which notes the various compound medicines mentioned in the works of Greek and Arabian writers. This work was written before 1300, for not only Peter de Abano, the celebrated counselor who died in 1316, but also Matthew Sylvaticus and Francis of Piedmont, physicians to Robert, the king of Sicily, quote from it in works written at the commencement of Robert's reign, which began in 1310. Thus, it is clear that Myrepsus must have lived before 1300. There is some evidence that Myrepsus was a

[1] Andronicus II succeeded the Emperor Alexins Ducas, who was surnamed Murzuphle, in the year 1204, and died in 1332, when he was succeeded by Andronicus the younger, upon whose death in 1341, Cantaeuzenus usurped the government under John Paleologus and retained it until 1355.

contemporary of Pope Nicholas the Third, who died in 1280 and was a great patron of learning. Therefore, Actuarius must have been a much older writer than Lambecius supposes, and, if we may be allowed to form any judgment from the superior purity of his style, he is older even than either Nonus or Psellus.

Actuarius's Compendium of Medicine

Regardless of the age in which he wrote and whether he preceded or followed Nonus and Psellus in the course of history, Actuarius certainly surpassed Nonus and Psellus in the classical purity of his works and the value of his observations. He practiced physic in Constantinople with much success. His six books *De Methodo Medendi* are said to have been written for one of the chief officers of the court[2]. In this work, he relies chiefly on Galen as his guide, along with Aetius and Paulus, although without naming either of them. However, he does not confine himself to these sources, but collects information from other authors. Actuarius deserves credit for introducing a multitude of original observations and noticing many things not found elsewhere. He named this compendium *Libellus*, and it was written for the private and exclusive use of the Ambassador, and not as a book for general practice. It does not discuss surgery or maladies that occur in females but does mention pediatric issues. Among other subjects, he particularly discusses aphthæ, or thrush, a complaint unknown except in infancy or old age. In the first two books, he discusses the causes and symptoms of diseases. In the next two books, he discusses their cures. The last two books contain a description of all the different remedies that he collected from the writings of the Greeks, from his own experience, and from other sources. In the third and fourth books of this compendium, and in other parts of his works, he discusses the results of his own practice at considerable length.

Hydrophobia and Palpitation

On the subject of hydrophobia, Actuarius notes intervals of six and twelve months between injury and the first appearance of symptoms. This observation has been confirmed by numerous cases in modern practice. He says that others, alluding to Paulus, speak of the virus remaining dormant for seven years. Morgagni even speaks of an interval of forty years. However, in such cases, we may reasonably doubt how far the disease is correctly referred to the cause, and whether it really originated from a more recent injury that has been overlooked or forgotten by the patient. Galen mentions a case in which an interval of a year elapsed from when the poison was communicated. Dioscorides remarks that although it usually exhibits itself before the fortieth day, it sometimes lies dormant for six or twelve months. Dioscorides, indeed, mentions cases in which an interval of seven years occurred, but he gives this simply as an instance not confirmed by any observation of his own. However, we may be inclined to question the fidelity of those accounts that speak of the hydrophobic virus lying dormant for

[2] Fabricius is guilty of an error in representing Actuarius himself as the Ambassador.

periods longer than twelve months, we must, from incontestable evidence, admit the perfect accuracy of what Actuarius provides as the result of his own personal knowledge.

Actuarius makes some new and valuable remarks about colic and hepatitis. The distinction he makes among the several causes of palpitation has every appearance of originality, whereas the whole of what Aetius and Orabasius have said upon the same subject is borrowed from Galen. Palpitation most frequently arises, according to Actuarius, from excessive heat or redundance of the blood, but sometimes results from vapors. The pulse is irregular in the former case, but regular in the latter case. None of the more modern of the ancient writers or those of the Middle Ages has accounted for this complaint in a more satisfactory manner. The Arabian writers who preceded or were contemporaries with Actuarius ascribed palpitation in general to cold. Paracelsus ascribed palpitation to the solution of buildup in the heart. Van Helmont ascribed palpitation to the acidity of the native gas. Sylvius de la Boe blames corrosive vapors arising from the pancreas as causing palpitation.

Actuarius says that inequality of the pulse indicates some obstruction in the heart, and forebodes palpitation, syncope, or sudden death.[3] In these cases, the pulse is unequal in the force of its strokes at intervals and also at times stops, for there is resistance from the blood either in the aorta or in the pulmonary artery. In order to overcome this resistance, the heart suspends its contraction until it has received a sufficient supply of spirits to propel the blood in the usual manner. Therefore, the intervals between the pulsations vary with the violence of the paroxysms, being longer or shorter in proportion to their severity. Such is the case when the complaint arises from plethora[4], and it is on this account that those who suffer from a suppression of hæmorrhoidal, menstrual, or any other habitual discharge are, as Galen observes, predisposed to palpitation. Palpitation may also result from excessive rarefaction[5] or from flatus pressing or distending either the thorax or the abdomen. Actuarius remarks that it is one of these etiologies that cause palpitation so frequently in hypochondriac men and hysteric women. Besides those remedies that are adapted to the nature of the case and the constitution of the patient, Actuarius chiefly relies on bleeding and purging to treat palpitation.

Smallpox and Purgatives

We unexpectedly find no discussion of smallpox in any works under the name of Actuarius. Although, had he lived as late as the period assigned to him by Lambecius, he must have known about smallpox, which had been introduced into Europe by the Saracens long before the fourteenth century. His silence on this subject leads us to conclude that he lived earlier than the time of Rhazes and the other Arabian writers who noticed it or that he was unacquainted with the works of these authors. Indeed, we do not read about any diseases in his works other than those diseases that had been already discussed by other Greek writers. It is slightly possible that Constantinople and the other parts of the Greek Empire in which Actuarius resided may not have been exposed to smallpox. Although the disease may have

[3] Galen predicted the death of the physician Antipatrus, who soon after expired, from this cause.
[4] Plethora refers to an excess of a bodily fluid, such as blood.
[5] Abnormal cohesion in the particles of the blood

evaded his actual observation, it is hardly possible that any report of its progress and manner of infection failed to reach him, or that he hesitated to discuss so novel and formidable a disease.

Actuarius is the first Greek writer who noticed the milder purgatives, such as Cassia[6], Manna, Senna, and Myrobalans. Of these, he says the last two were imported from Syria and Egypt. He describes Senna as a fruit and does not take notice of the leaves, which are the only part employed in modern practice. He acknowledges when he borrows information on these purgatives from the Arabian writers. Like the Arabian writers, he describes three kinds of Myrobalans. He devotes an entire chapter to syrups and juleps, preparations that he borrowed from the Arabians. He also discusses distilled waters, a subject not noticed by any of the Greek writers who preceded him.

Other Works of Actuarius

Actuarius wrote two books on the animal spirits in which his reasoning concerning physiology rests upon principles taken from Aristotle, Galen, and others who have discussed the same subject. He also wrote seven books on urine. He discussed this subject in such great detail that he hardly left room for later writers to add more information, and many of these writers have transcribed his commentary almost word for word without acknowledging him.

Nonus

Lambecius appears to have assigned Nonus a date earlier than that to which he is entitled. Lambecius has placed Nonus in the reign of Constantine the Seventh, who was surnamed Porphyrogenitus and died, according to Lambecius, in the year 900. However, according to Maclaine in his translation of Mosheim's *Ecclesiastical History*, Constantine the Seventh died in the year 959. He bases this opinion on the fact that Nonus dedicated his work *Enchiridion Medicum* to a Constantine Porphyrogenitus, whom Jeremias Martius, who edited this work in Greek and Latin, thinks may have been the son of Constantine Ducas, or Constantine the Tenth, who died in 1067.

The *Epitome* of Nonus is a mere compilation. For instance, the whole of what he says in his chapter on Carus about the anterior part of the brain is borrowed from Alexander and Paulus; specifically, his directions concerning bleeding in the paroxysm of calculus are from Paulus, and the distinction he makes between the effects of bleeding and purging in pleurisy are from Alexander. Even his remedies are copied from Aetius. However, he did not acknowledge these authorities and even claims these experiences as his own. This work is ascribed to Theophanes without the slightest mention of Nonus. It is said in the title to have been compiled chiefly from the work of Oribasius. However, it is evident that the compiler,

[6] The pulp of the fruit of the Cathartocarpus fistula, a tree originally introduced into the West from the East Indies and now frequently to be found in the Carribean islands, especially in Dominica, where it makes a most splendid appearance when in flower in the months of May, June, and July, rivaling the Laburnum of Europe in the elegance of its pendulous racemes of golden flowers.

whether Nonus or Theophanes, borrowed much more from other sources and hardly from Oribasius.

Psellus

Michael Psellus was nearly contemporary with Nonus. Psellus compiled a work on the nature and properties of aliments and dedicated it to the Emperor Constantine, who reigned from 1043 to 1055 and whose surname, as Lambecius conjectures, was Monomachus. However, if Psellus died in 1078, as Lambecius himself admits, the individual to whom the work was dedicated is probably Emperor Constantine Ducas. This conclusion is more likely because it appears, per the testimony of Zonara, that Psellus was a private tutor to Michael Ducas, the son of Constantine Ducas. We learn also from Zonara that he was perfectly illiterate and utterly unfit for such a position, whereas Anna Comnena speaks of him as a man of vast genius and profound erudition. Leo Allatius praises him equally and, in his dissertation on the Pselli, ranks him among the first writers of the age. Thus, there may have been two individuals named Michael Psellus, to each of whom these contradictory characterizations applied. The aforementioned work on aliments is a mere compilation from the older Greek writers, and it is probable that the Psellus spoken of by Zonara authored it. The Psellus eulogized by Allatius wrote many works, but was so harassed and plundered by Nicephoras Botonias that he retired to a monastery, where he died at an advanced age.

Although it was a compilation, the work of Psellus furnished material for a book by Simeon of Antioch. As Psellus's book was well known and published recently, it is surprising Simeon pilfered so freely from its contents. Although junior to Psellus, Simeon must have been nearly his contemporary, for he dedicated his work to Michael Ducas, surnamed Parapinaceus, who abdicated the Empire in 1078, the same year in which, according to Lambecius, Psellus died. Several other works by Simeon still exist, including a Greek translation of an Arabic work on Indian philosophy composed at the request of Chosroes, king of Persia, by the physician Perzoe.

Demetrius Pepagomenus

Our only clue as to when Demetrius Pepagomenus lived is the dedication of his treatise on gout to Michael Palæologus. We are unsure as to which of the two emperors who shared that name the dedication belongs. The first succeeded to the throne in 1260, and the second in 1355.[7] Demetrius's treatise on gout, although chiefly a compilation of works by Alexander and others, is still above mediocrity. Fabricius seems to think that the treatise on the cure for calculus, which is erroneously ascribed to Galen, was actually developed by Demetrius, but he gives no reason for this opinion.

[7] Friend says, 1310 (Hist. Med. p. 208) but I have preferred Maclaine's tables, in which Andronicus II is said to have succeeded Michael Palæologus in 1283 and continued to reign till 1352, when he was succeeded by Andronicus the Younger, on whose death John Cantaeuzenus usurped the government till 1355, when John Palæologus recovered the throne and reigned under the title of John VI until 1390.

Chapter 6

The Fetters of Corruption and the Progress of Medical Education

Abuse and Corruption of Medical Practice by Religious Authorities

It has been already mentioned in the fourth chapter that the Arabs blindly followed the ancient fathers of Greek literature and philosophy, such as Hippocrates, Aristotle, and Galen. However, literature and medical science are more indebted to the Arabs during the period following the overthrow of the Roman Empire than to those men who, in the earlier ages, corrupted the Christian faith and enveloped science in superstition. Although the Arab Muslims did not critically evaluate and improve upon the teachings of the Greeks, they preserved, and did not debase, those teachings. Like the pagan priests prior to the reformation of medical science promoted by Hippocrates, the earlier clergy of the Christians claimed the practice of medicine as their exclusive privilege and disgraced it by implementing a system of charlatanry.

The monks and priests of those ages rushed into the practice of medicine without the slightest preliminary education. They were indifferent to the success of their practice as long as they pocketed its fees and careless about their patients as long as they could satisfy their own greed. The practice of medicine enabled monks to earn money at the expense of the laity by enabling them to add the terrors of the church to the authority of the physician, and gave them access to more victims when the agony of disease made them less reasonable and thus more likely to be fooled. In the event that a patient became so ill or presented with such novel symptoms that their limited skills could not address the matter, the monks hid their deficiency in knowledge beneath the veil of mystery and called upon the power of faith.

Accordingly, the first Lateran Council, which was held in 1123 during the reign of Pope Calistus the Second, prohibited the interference of the regular clergy in the practice of medicine and their attendance at the bedside of patients other than as consoling ministers. This prohibition was found to be inadequate, and a second prohibition was promulgated by the Council of Rheims in 1131 during the reign of Pope Innocent the Second. A third prohibition, conceived in terms stronger than either of the two preceding it, was passed by the second Lateran Council in 1139, and threatened with the severest penalties those monks and canons who applied to the practice of medicine, "neglecting the sacred objects of their own profession, and holding out the delusive hope of health in exchange for ungodly lucre." It ordained that all bishops, abbots, and priors who connived at such proceedings of the clergy within their respective jurisdictions should be suspended from their ecclesiastical functions.

However, the clergy tended to disregard the decrees of councils in those days, and they continued to be rebellious towards the authority that they had sworn to obey. In France, the stubbornness of the clergy rose to the highest pitch, and their obstinacy made the longest stand. In Britain, on the contrary, these abuses appear to have been confined within straighter limits, and the character of the clergy to have assumed a higher tone. As early as the seventh and eighth centuries, they could boast among their numbers men of the highest literary achievements and the most ardent Christian piety. They were esteemed for the superiority of their literary and scientific achievements and the comparative purity of their morals. Thus, they were in the highest demand among the princes of other countries and were eagerly sought after to fill positions of the highest importance.

Alcuinus

Among the learned men who adorned the court of the illustrious Charlemagne, the clergy of Britain occupied the most distinguished station. None among them was more recognized for talent and more respected for virtue than Alcuinus, an abbot of Canterbury, who in 790 was sent by Offa, king of Mercia, on a mission to Charlemagne. He joined Charlemagne's court and became his instructor in science. The University of Paris owes its origin to Alcuinus, who also instigated the establishment of similar institutions in France and Italy. Although a churchman, Alcuinus does not appear to have been less learned in the science of medicine than in theology. The following lines from one of his poems evidence that both the study and the practice of medicine were familiar subjects for the members of the Imperial Academy, which was the foundation of the university that was afterwards established,

> Next Physic's sons their useful labors ply,
> And all their varied arts of healing try,
> Life's sanguine stream here one commands to flow;
> The ample bowl another holds below;
> Another here each active simple blends;
> The gruel there another's skill attends;
> And one, again, the soothing draught prepares,
> Of pow'r to blunt each pang of mortal cares.

At length, Alcuinus left Charlemagne's court and retired to the abbey of St. Martin's in Tours, from where he maintained a regular correspondence with Charlemagne for the remainder of his life. Alcuinus died at the abbey of St. Martin's in 804.

The School of Salernum

The clergy of France stubbornly continued to practice medicine and defied the commands of the Pope and the decrees of the Councils. After three centuries, common sense and a regard for the public good finally prevailed. It was only by a special decree, which the Cardinal d'Estonteville procured from the Pope, that physicians were permitted to marry. This helped divorce medicine from theology, thereby ending the shameful impostures and superstitious practices that were prevalent.

During this dark time of clerical ignorance, imposture, and presumption, the School of Salernum[1] became well known. It originated from a Benedictine monastery established in Salernum during an early period of Christian history, the monks of which had for many centuries practiced medicine in the ignorant and superstitious manner characteristic of the pagan priests before the days of Hippocrates. Around the ninth and tenth centuries, the monks began to study regularly the best medical writers among both the Greeks and the Arabs, chiefly taking as their guides the works of Galen, whom they highly regarded. Thus, they gradually emancipated medicine from the constraints in which it had been held for so long, raised it to its proper rank among the sciences, and obtained for themselves wealth and reputation through the celebrity of the pupils whom they instructed.

Constantine the African

Constantine the African was born in Carthage around 1010 and distinguished himself with his proficiency in medicine at the School of Salernum. Knowledge of the Greek and Arabian systems of medicine were introduced into Italy through him. Constantine travelled in his early youth to the East and resided in Babylon and Bagdad over a span of 30 years, during which he studied Oriental thought and languages. He returned to Carthage, but, finding his life in danger from the malignity of its inhabitants, left and took refuge in Apulia in Italy around 1060. He brought with him letters of recommendation to Robert Guiscard, the reigning Duke, from whom he received the appointment of secretary, which, as he was obliged to reside at the town of Reggio in Calabria, obtained for him the name of Reghinus. He entered the Benedictine monastery on Mount Casinus, where he died in 1087.

Constantine authored many works. He used the materials left by the ancients for his writing, but his writing is not devoid of originality. He translated Isaac's book on fevers from Arabic into Latin and also authored some Greek translations. He claimed to be the first who wrote expressly about disorders of the stomach. This work, which he dedicated to Alfanus Primus,[2] exhibits almost every important fact connected with this subject found scattered among the volumes of the ancients. In this work, he repeatedly mentions John of Damascus,[3] who appears to have been altogether distinct from the Mesue of the Arabs, for he speaks of

---

[1] Hamilton writes that Salernum is a town situated within the territories of Naples.

[2] Alfanus was Archbishop of Salernum from 1057 to 1087 and known for studying both general literature and medicine.

[3] That John of Damascus could not be the same with the Arabian writer Mesue, who flourished in Baghdad around the year 865 in the time of the Caliph Haroun Al Raschid, is evident from his noticing Avenzoar, who did not flourish before the end of the tenth or beginning of the eleventh century. Thus, he could not have lived before the middle or close of the eleventh century and must have been a contemporary of Constantine.

medicines described by him of which not the slightest mention is to be found in the works of Mesue.

Constantine also developed a distinct theory on melancholy, which is thought by many to be little more than a transcript of the work of Rufus the Ephesian, which is now unfortunately lost but was extant when Constantine was writing. He also composed a work that he termed his "commonplace book", a collection of all the important facts concerning the theory and practice of medicine compiled chiefly from the works of the Greek and Latin writers. He states that he produced this work because at the time, there was no similar compilation that was properly written. However, this compilation is little more than a translation of the *Almaleci* of Haly Abbas, unaltered even with regards to the number and divisions of its books and chapters. The writings of the Arabs during this time were unknown in Italy, and the writings of the Greeks were nearly extinct. Thus, Constantine ran little risk of being detected as a plagiarist. In his compilation, he does not mention the name of Haly Abbas, Isaac, or a single Arabian writer even once. To note, Marcellus Empiricus plagiarized from the writings of Scribonius Largus. However, Constantine was a zealous promoter of medical science, and through his influence Duke Robert extended patronage to the School of Salernum. After the death of Constantine, the School of Salernum rapidly rose in status. During the period of the Crusades,[4] the town of Salernum became a place of great resort for the multitude of warriors who were traveling back and forth between Europe and Palestine, and attained celebrity in the early twelfth century.

Robert, Duke of Normandy

Among other distinguished persons who visited Salernum was Robert, Duke of Normandy and son of William the Conqueror. He arrived at Salernum around 1100 seeking medical advice after having participated in the Crusades and having been wounded there in the arm with a poisoned arrow. He was accompanied by his wife Sybilla, daughter of Galfridus, Count of Conversana, and for her sake, Robert had sacrificed his chance of succeeding to the throne of England upon the death of his brother William Rufus by living with her in Italy. As a result of neglect, Robert's wound had turned into a fistulous ulcer. The medical practitioners of Salernum decided that the only means of extracting the poison that was preventing the wound from healing was by suction. However, Robert refused to listen to this proposal, as such a procedure would endanger the operator, but his wife Sibylla, upon hearing the advice of the physicians, extracted the poison from his wound with her own mouth while he was in an opiate-induced sleep, but at the cost of her own life. Robert's visit to Salernum is commemorated in *Regimen Sanitatis Salemi*, a poetic treatise on the art of preserving health, by John of Milan. This work contained a chapter on the treatment of fistulous ulcers as a result of Robert having endured such a complication.

[4] It was in the year 1094 that Peter the Hermit, a native of Amiens in Picardy, first stirred up the princes and people of Europe to attempt the recovery of Palestine from the dominion of the Saracens, and thus occasioned those wars that took the name of "Crusades" from the white cross worn as a badge by all those engaged in them.

Ægidius

Ægidius[5] was a native of Athens and a monk of the order of St. Benedict who was appointed first physician to Philip Augustus in the latter part of the twelfth century. Ægidius, in imitation of John of Milan, wrote a poetic treatise on the properties of medicines and on the pulse and urine. In this work, he finds fault with Constantine for having written too diffusely and with Philaretus for having written too concisely with regards to the pulse and urine. He also mentions those who were educated at Montpellier in France, which during this time was celebrated for its medical school. Like its predecessor, this poem became so popular that it became a textbook used in the schools, and Gentile, one of the ablest commentators of the age, expounded it.

Benjamin

Duke Roger, the first King of both the Sicilies in the year 1130, and his successors William the First and Second promoted the improvement of medical knowledge at Salernum such that Orderic Vital, the historian who flourished at the time and died in 1141, predicted that the School of Salernum would become famous throughout the world. Indeed, Benjamin of Tudela[6] praised it as the best school of medicine among the children of Edom.[7] He visited Salernum around 1172 on his return from travels over nearly the whole of the then-known world.

In the itinerary of his travels, Benjamin[8] enumerates the several cities that were inhabited by the Jews at that period and adds that many of them pursued the practice of medicine not merely among their own tribes, but also among the Moors and Christians, despite the express prohibition by the canon law of that period against Jews practicing medicine among Christians. However, laws are powerless when not in conformity with public opinion, and the laws enacted to discourage the Jews were only efficient as long as the Christians maintained something like an equality of talent, wealth, and learning with their Hebrew brethren.

---

[5] Besides this Ægidius, we have some account of another of the same name, who was surnamed Atheniensis and flourished as a physician and philosopher under Emperor Tiberius II who died in 586. This Ægidius also became a Benedictine and was the author of several tracts, of which the principal ones are *De Pulsibus* and *De Venenis*. Being accidentally wounded by an arrow, he would not allow the wound to be dressed so that he might exercise his fortitude in bearing pain. The former Ægidius was a canon of Paris and named Ægidius Corbolensis, or Gilles de Corbeil. His work on the urine and pulse with an explanation and commentary by M. Gentilis de Fuligineo was printed in Venice in 1494, in Lyons in 1505, and in Basil in 1579.

[6] Hamilton writes that Tudela is a city of Navarre in Spain, and that it is situated on the right bank of the Ebro about 23 leagues northwest of Saragosa.

[7] Edom is the name by which Benjamin of Tudela distinguishes the Christians of the West.

[8] Benjamin was a Jewish rabbi. He was born in Tudela about the middle of the twelfth century. Being anxious to visit his countrymen in the East, whom he hoped to find in such a state of power and opulence as might redound to the honor of his nation, he commenced his travels in 1160, proceeding from Spain to Constantinople and travelling as far as China. He then returned through various parts of India, and embarked on the Indian Ocean, visited several of its islands, and reached Europe by way of Egypt after an absence of 13 years. He died in 1173, a few months after his return. His itinerary, the first edition of which appeared in Constantinople in 1543 with a Latin translation by Benedict Arias Montanus, contains an account of his travels, intermixed with statements, no doubt somewhat exaggerated, of the power and prosperity of his Jewish brethren. Caspar Oudin, however, states that he had a most religious regard for truth, and that his observations and accounts have proved upon examination to be generally exact. See, *Oud. Comm. de Scrip. Eccles. Tom.* ii. *col.* 1524. *Lipsiæ* 1722.

Therefore, no sooner did the balance decidedly preponderate in favor of Jewish over Christian learning and talents than the laws became virtually a dead letter and popular prejudice yielded to the force of private interest. Jews were openly employed and publicly appointed physicians to the first rulers of the day. Thus, we find almost every ruler in Christendom, both before and during the period in which Benjamin flourished, employing Jewish physicians and reimbursing them with princely salaries. For example, Charlemagne employed two Jews as personal physicians, Farragutus and Buhahyliha Bengesta, who wrote at his request the work *Tacuin* or *Tables of Health*.

Toward the end of the tenth century, Jews held the first rank throughout Europe as physicians and were employed professionally by the popes themselves, as they were one of the few groups who understood the teachings of the Arabs. Their superior knowledge gained them advantages among the Saracens in Spain, even though the Moors were more prejudiced against them than the Christians. Joining the Moors in their invasion of Spain and expulsion of the Christians from Spain about the year 714, the Jews obtained a settlement in the cities of Cordova and Grenada. The Jews are said to have had an academy or university at Sora in Asia as early as 204, and to have enjoyed the confidence of many of the earlier caliphs by whom they were employed. From this short review of the general state of learning among the Jews, both at the period and for many centuries before Benjamin flourished, we can somewhat gauge his competence at making statements about the merits of the School of Salernum, especially when his conclusions were corroborated by other writers of the most unquestionable authority.

Frederick II

Frederick II, Emperor of the West, endowed the School of Salernum with certain privileges[9] in the year 1225. At the same time, he established a curriculum for the school's pupils. He also regulated the several examinations that the candidates for degrees were to undergo before being licensed to practice either in medicine or surgery. Frederick II's policies, together with the privileges he granted, rapidly raised the reputation of the School of Salernum that it retained for many years.

Policies and Decline of the School of Salernum

The statutes of Salernum are of extreme antiquity, but admirably adapted to the object which they had in view and devised with no inconsiderable degree of ability. Being the first of their kind, they served as a model for laws subsequently framed for the government and discipline of similar establishments. The school was placed under the protection of St. Matthew and was styled, according to the inscription on its seal, *Civitas Hippocratis*, or the City of Hippocrates. The foundation consisted of 10 doctors or professors whose seniority was regulated by the dates of their appointments. The examination for degrees was conducted

[9] Among the privileges, or rather monopolies, granted by the Emperor at this time, it was not perhaps one of the least important, that the privilege of granting degrees and a license to practice was restricted to the schools of Salernum and Naples.

with the greatest strictness, and the works on which the candidates were tested were *Therapeutics* by Galen, the beginning of the *First Canon* by Avicenna, or *Aphorisms* by Hippocrates. Candidates for the degree of Doctor of Medicine were required to be 21 years old and to produce testimonials of having studied medicine under competent professors for seven years. To become a surgeon, it was necessary to have studied anatomy for at least one year. On being admitted to the school, the candidate was required to pledge compliance with the laws and usages of the school, to refuse all remuneration for attending to the poor, and not to enter into any lucrative contract with a pharmacist. Having sworn faithfully to observe all these regulations, a book was placed in his hand, a ring upon his finger, and a laurel crown upon his head, and he was then dismissed with a kiss of peace. In addition to the statutes concerning admission and the curriculum, there were many other provisions for regulating the practice of medicine within the jurisdiction of the school, more especially for regulating the department of pharmacy. The pharmacists were obliged not only to compound their medicines faithfully according to the prescriptions of physicians, but also to sell them at a price regulated by a competent authority.

The various professors at the School of Salernum were men distinguished for their learning, talents, and assiduity. Under the auspices of Frederick II, they produced copies of the works of the Arabs and published translations of those works in Greek and Latin. Although these endeavors caused the original writings of the Greek physicians to fall into obscurity for two or three centuries, the translations of the medical writings of the Arabs kept alive the recollections of the Greek writers and eventually led to their being rescued from destruction. The School of Salernum boasted the presence of Enos, Gariopontus, Nicholaus, and others who were known as academic prodigies. However, the school was doomed to fall into decay. It exhibited a tarnished splendor at the start of the fourteenth century, and toward the middle of that century had declined so far that Petrarch made the following observation, Report speaks of Salerno as having been a fountain of medical knowledge, but there is nothing which does not become dried up by age.

Chapter 7

Medicine from the Tenth to the Sixteenth Centuries

Albertus Magnus

In the universal ignorance that enveloped the whole Christian world, pre-eminence in knowledge often became dangerous to possess. Several of the Popes were regarded during the eleventh and succeeding centuries as magicians solely on account of their attainments in mathematics and philosophy. To the ignorant, the book of nature and the records of inspiration were as a sealed volume, to unclose which was an act of the deepest impiety, and men were taught to view knowledge as a kind of Pandora's box, which, if opened, would let loose an inundation of misery and pestilence. Yet, in these ages of darkness, among the monks and clergy, some few meteors shone out with a brilliancy that for a time promised to dissipate the surrounding darkness.

The celebrated Albertus Magnus was born in the fifth year of the thirteenth century at Lawingen in Suabia and educated at the University of Pavia, from where he proceeded to Paris, where he took the degree of Doctor of Medicine in 1236. Happening to hear the celebrated orator Father Jourdain preach, he was induced to assume the habit of the Dominicans. Nevertheless, he became a popular lecturer in philosophy, medicine, and theology both in Paris and Cologne, and boasted among his pupils the celebrated Thomas Aquinas. On the death of Father Jourdain, he was promoted to the rank of Vicar General, and afterwards, to Provincial of the Dominican Order. In 1260, he was raised to the seat of Ratisbon, which he retained for three years, and then retired to Cologne. From there, he proceeded on a missionary tour through Germany and Bohemia to preach a fourth Crusade against the Turks. In 1274, we find him assisting at a council of the Church held at the city of Lyons. From this time to his death in November 1280, he resided almost wholly at Cologne, where he occupied himself as a religious instructor of the Dominican Order.

He was a voluminous writer upon a vast variety of subjects. Several works that have been imputed to him are undoubtedly spurious, especially the theory *De secretis mulierum, item de virtutibus herbarum, lapidum, et animalium*, which was the production of his pupil Henry of Saxony. Albertus is said to have devoted much of his time to the study of chemistry, motivated by the search after the philosopher's stone and the elixir of life. His proficiency in this science, as well as in mechanics and other branches of general knowledge, subjected him to the imputation of magic despite his belonging to the Dominican Order, who wielded at that time, the thunders of the Inquisition. He constructed an androides, or automated figure, which the ignorant multitude believed to be a familiar spirit by whom he was assisted in his supposed conjurations. This figure, we are told, was so artfully formed as to be able to speak and perform a number of other feats that may well have passed for the supernatural in those days.

Some have given Albertus credit for the invention of firearms without perhaps recollecting that the discovery of gunpowder, generally ascribed to Bacon at a considerably later period, ought to take the precedence of firearms, which without gunpowder would have been of little practical utility. Before the days of Albertus Magnus and Bacon, the knowledge of chemistry, or, more correctly, alchemy, was restricted almost wholly to the Arabs. However, this fascinating science now was communicated to the inhabitants of Germany and England through the exertions of Albertus in the one and Roger Bacon in the other, who, along with Peter of Maharn Court in Picardy, were the only three persons in Europe possessing the slightest acquaintance with this art.

Bacon

Roger Bacon was born at Ilchester, Somerset, in 1214, and received the first part of his education at Oxford, from where he left for the University of Paris, where, in testimony of his superiority in all the literary and scientific attainments of the age, he was promoted early to the dignity of Doctor in Theology. He assumed the monastic habit of the Order of St. Francis and returned in 1240 to his birthplace. On his return from France, so that he might be enabled to study with more advantage and less danger of interruption, Bacon established himself at the University of Oxford.

Such was the esteem in which Bacon was held that he was enabled, by the assistance of liberal contributions, to accumulate a library of the most valuable works, supply himself with the most extensive apparatus, and conduct diversified experiments. As might have been expected, the growing reputation of the sage awakened in time the envy of his brethren of the Franciscan Order, who anticipated, in the increasing illumination to which his lectures gave birth, the approaching exposure of their own system of imposture. Therefore, to begin their persecution, they circulated slander whose baseness was only equaled by their falsehood, and exhibited his scientific experiments as effected through the operations of magic and the agency of the spirits of darkness. Having thus so far prevailed with the stale pretense that the innovations that Bacon had introduced into the doctrines of science were calculated to disturb the tranquility of the Church, they sought to obtain an order prohibiting the continuance of his lectures. Not satisfied with thus drying up the fountain of knowledge at its source, they at length confined him so closely as to deprive him even of the cheering consolations of friendship and prevent his receiving the supply of food required for the sustenance of exhausted nature.

The prelates and the monks, as he complains himself in his letter to his enlightened and open-minded friend Pope Clement IV, were apprehensive his writings might extend beyond the putrefying atmosphere of their convent. However, had the cause of religion alone been in danger, Bacon might have continued to lecture unmolested by them. His crime was held to be of a deeper dye than a mere attack upon Christianity. He dared to reprobate the ignorance and licentiousness of the priests. Those men were sufficiently sensitive when their own abominations were held up to public criticism and contempt. Another crime of Bacon's, unpardonable in the eyes of the priesthood, was his intimacy with Robert Greathead, the virtuous and talented Bishop of Lincoln, who addressed a letter of salutary reproof to Pope Innocent IV and did not hesitate to express his unqualified belief in the identity of the Pope with the Antichrist, in addition to which Bacon had been guilty of the same offence as had been committed by his friend the Bishop, having also written freely to the Pope on the subject of reformation in the Church.

Despite the slandering by the monks, such was the opinion entertained of Bacon by the Cardinal Bishop of Sabina, who was at this period resident as Legate from the Pope in England, that this prelate requested Bacon to furnish him with a complete collection of his numerous and important writings. However, apprehensive of the consequences of openly violating the prohibition of his fraternity, which peremptorily forbade his communicating copies of his works to any person whatsoever, Bacon at first refused compliance with the wish of this exemplary Bishop. When, however, on the death of Urban IV in 1264, the virtuous Bishop of Sabina was elected to the vacant chair under the title of Pope Clement IV, Bacon lost no time in acquainting him with his readiness to comply with the wishes he had formerly expressed, provided his compliance would be unattended with danger to himself. In reply, Clement having assured him of his protection against any interference of his own order, began to collect, arrange, and improve his several works, and, having digested the whole into one volume under the title of *Opus majus*, he sent it to the Pope in 1267 by John of Paris, a favorite pupil. The Pope was so grateful for this present that he afforded Bacon in return every encouragement and facility in the conduction of his studies, which his ample power enabled him to bestow.

Unfortunately for Bacon, the reign of this virtuous Pope drew to a rapid close, and his death in the following year transferred the scepter of power from the hands of an indulgent friend to those of a remorseless and implacable foe. After the death of Clement, his old persecutor and vindictive adversary, Jerome de Esculo, or Ascoli, the General of the Franciscans, obtained the full sanction of Pope Nicholas III to imprison him again and prohibit his works from being read on the false and frivolous pretext of their theories on alchemy, astrology, magic, and necromancy.

With an equally shortsighted wisdom, Jerome, wielding the delegated authority of Nicholas and his successors, detained the unfortunate victim of his fears and animosity upwards of 10 years in confinement. On his succeeding at length to the Pontificate himself upon the death of Honorius IV in 1288, Bacon made a fruitless attempt to conciliate his favor and procure indulgence by writing and transmitting to him a copy of his treatise on old age with the means of mitigating its infirmities and averting its evils, nor did he obtain his freedom till near the close of his persecutor's tyrannical existence in 1292, when the appeal of his liberation was at length granted to the entreaties of an English nobleman. Bacon returned once more to the banks of the Isis and composed at Oxford a compendium of theology. This was the last of his productions and appears, from internal evidence, to have been composed sometime in the year 1291. In the year immediately succeeding, he died on June 11 at the age of 78 years.

Bacon's attainments were prodigious for the age in which he lived, and the versatility of his genius is truly surprising. He mastered Greek, Hebrew, and Latin, and in mathematics he is represented by some of his admirers as superior to Archimedes himself. His works, which are numerous almost beyond example, contain the latent germs of at least half of the most important of our modern discoveries and improvements, which his genius enabled him to foresee and would have enabled him to realize had he lived in a more favorable age. His acquaintance with optics was such that persons have been awarding him the credit of inventing the telescope. In geography, his research was various and extensive. In astronomy, his penetrating judgment enabled him to detect the errors that had crept into the calendar and point out the remedy that was adopted, nearly three centuries after his death, by Pope Gregory XIII. He wrote a total of 99 different works that embraced nearly the whole circle of human knowledge.

The prevailing rage among the men of science in Bacon's day was the discovery of the philosopher's stone, which was to convert all the inferior metals into gold, and the composition of the elixir of life, which was to confer perpetual youth and unfailing vigor upon man. However foolish these pursuits may appear, the substantial advantages that science has derived from their pursuit and the important discoveries that have resulted from the vain attempts to attain these unattainable objects have been of such magnitude that we should view with gratitude this foolishness. That a mind like Bacon's participated in these endeavors to some degree furnishes only an additional proof of that weakness of our nature, which brings the learned and the unlearned, the wise and the simple, in some respects, so nearly to the same level.

Medicine at European Universities

Medicine had now become a general branch of education at most European universities, but the doctrines of Hippocrates and Galen were adhered to with such slavish servility and the science was so tainted by the scholastic philosophy of the period that little improvement was made. The universities with the highest reputation as schools of medicine were those of Montpelier, Paris, Bologna, Padua, Ferrara, Pavia, Milan, and Piacenza. In 1271, the surgeon Pitard laid the foundation of the College of Surgeons in Paris, an event productive of beneficial results to this department of medical science, which became from this period successfully practiced in France as a distinct branch of the medical profession.

Petrus de Apono

At the same time as Bacon flourished, Petrus de Apono prospered near Padua where he was born in 1253. He was distinguished for his skill in chemistry as well his knowledge of medicine. He went at an early age to Constantinople for the purpose of learning Greek, and then to Paris to study medicine and mathematics. About this time, he likely visited England. Upon his return to Italy, he was invited to fill a professor's chair at Padua, from where he later moved to Bologna. On being sent for to attend Pope Honorius IV, who died in 1288, he stipulated for the daily fee of 400 ducats as long as he should be detained in Rome.

Unfortunately for himself, de Apono pretended to an acquaintance with astrology and to foretell, by its aid, the termination of diseases. By this idle boast, he subjected himself to the charge of magic and awakened the vengeance of the Inquisition. Of this charge he was acquitted. However, upon a repetition of the accusation around 1319, he was not so fortunate, and death alone rescued him, at the age of 66 years, from the tender mercies of the Inquisition. Sentence was pronounced against him for contumacy, and his body ordered to be disinterred for the purpose of expiating on the pile the pretended heresy of the soul. His friends, however, having privately removed the body to another place of interment, the Holy Office was forced to be content with wreaking its vengeance upon his image.

As a medical writer, the opinions given by authors respecting de Apono are various and contradictory. Bernardus Scardeonius and Naudaeus praise him, while Champerius pronounces him to have been a man of much reading but little judgment. We shall most probably find the just estimate of his merits to lie midway between these extremes. During his residence in Paris, sometime between 1260 and 1270, he composed a work entitled *Conciliator differentiarum philosophorum, ac preecipue medicorum*, which obtained for him great applause and the title of "Conciliator". The object of this work, which he afterwards dedicated to his great friend Pope John the XXI, appears to have been to reconcile the conflicting opinions of the philosophers and physicians, both of his own times and of antiquity. He also wrote an essay on poisons and their antidotes.

Gentilis, Arnoldus de Villa Nova, and Lully

Synchronous with Bacon and de Apono flourished Gentilis, who was born at Foligni around 1230 and studied medicine under Thaddæus of Florence. He was regarded upon his return to Foligni as one of the prodigies of the age and was reputed to be one of the best commentators on Avicenna. Gentilis died at Bologna around 1310, leaving behind him several papers that were collected and published in Venice in 1484, 1486, and 1492.

Also at the time of Bacon, de Apono, and Gentilis flourished Arnoldus de Villa Nova, or Arnaud of Villeneuve, where he was born about the year 1250. Having completed his elementary studies at Paris and Montpellier and educating himself further by visiting different schools in Italy, he went to Spain to learn the teachings and language of the Arabs and profit from their acquaintance with the art of healing. While in Spain, he formed an acquaintance with the celebrated Raymond Lully and acquired so high a reputation that a sect was formed in that country that assumed the title of Arnoldistæ. De Villa Nova distinguished himself mainly by his investigations in chemistry, on which he wrote many works. In his abridgement of the practice of medicine, he describes a number of chemical medicines. Richard of England, in his chemical essay entitled *Correctorium*, states that Pope Innocent V, when suffering from an attack of plague, was cured by de Villa Nova, who administered to him the tincture of gold. Arnaud, in his abridgement, has listed formulas collected not only from the works of his predecessors, but also from his contemporaries. His studies extended to theology and medicine, and he bore the character of an able adversary. He had such implicit faith in astrology that he confidently predicted the destruction of the world in the year 1376. When in Paris, he offended the priests by the freedom of his theological opinions, those especially related to the immoral and scandalous lives of the monks. The faculty of theology in Paris condemned several of his positions.

To escape persecution, especially as the Inquisition had already commenced proceedings against Petrus de Apono, he retired to the court of Frederick of Arragon, with whom he had a friendship during his former residence in Spain. Here he wrote his essay on the government of the health and his commentaries on the School of Salernum. He was also the author of many other works. In these, he complains much of the interference of the clergy in the practice of healing, an evil that continued perpetually to increase despite the repeated efforts of the Popes to check it, until after the revival of letters. Among other works of his we find *Expositiones visionum quae sunt in somniis ad utilitatem medicinae*, *Remedia contra veneficia*, and *De conservanda juventute et retardanda senectute*, which strongly demonstrates the superstitious bent of his mind. The period of his death is not known with certainty, but it is presumed to have taken place around 1312, for we find Pope Clement V in 1313 sending letters asking all who received them to send to him, if they happened to possess it, a compilation on the practice of medicine that de Villa Nova had promised to send him.

Raymond Lully was born in 1236 and with de Villa Nova and Thaddæus of Florence, who was born in 1270 and is the first who makes mention of spirits of wine. Lully regarded this as an emanation of the deity, wisely concealed from the earlier inhabitants of the earth and reserved for the renovation of the old age of the world. Therefore, the fact of its discovery led him to argue that the end of the world was rapidly approaching. It is in the works of this writer, who died around 1315, that we find the term "alcohol", though evidently of Arabic origin, first applied to spirit of wine.

Brunus and Theodoric

About this period also flourished Brunus, an Italian physician and surgeon who was born in Calabria and is said to have been intimate with Petrarch. He published at Padua in 1252 a work on surgery more copious than any which preceded it, but chiefly taken, as he himself acknowledges, from Arabian writers, particularly Albucasis, from whose works he has taken the account of the operation of lithotomy. Not long after Brunus died, Theodoric, Bishop of Cervia, endeavored to acquire a reputation for himself at Brunus's expense by publishing his work on surgery in his own name.

William of Salicetum

About the same period lived William of Salicetum, who was a professor at Verona and died, as Lindanus informs us, in 1270 or 1280. Despite the freedom with which he borrowed from the works of the Arabs and others who preceded him, the productions of William possess considerable merit and display a larger proportion of original facts and observations than was usual in his day. Indeed, he appears to have been superior to all his contemporaries in talent and erudition both in medicine and surgery. His practice seems to have been founded on the basis of long and careful observation. He assumed it as an axiom that medical knowledge was not to be acquired from the study of books, but from practical experience. He appears to have been the first who recommended mercurial lotions for the face.

Lanfranc, Gordonius, and Mundinus

Lanfranc, who was born at Milan, carried out his studies chiefly at Leyden, from where he went to Paris in 1295. In Paris the following year, he completed his work on medicine, nearly the whole of which is taken from Gulielmus de Saliceto with merely a rearrangement and to whom he has not made the slightest acknowledgment. His opinions upon some points are unique, however. Gordonius was a professor of medicine at Montpellier beginning in 1285, and he taught there for more than 20 years. He was the author of several medical works, which were collected together and published at Ferrara in 1487. About the same period with Gordonius, flourished Mondino or Mundinus, who was born in Milan around 1276 and came to hold the professorship of medicine at Bologna in 1316. Mundinus, who enjoyed a great reputation for professional skill throughout Italy, was chiefly remarkable for his efforts to improve the languishing science of anatomy through dissection. Around 1315, he composed a work on anatomy interspersed with various original observations and discoveries, with these discoveries related chiefly to the uterus.

Sylvaticus and Gilbertus Anglicanus

Nearly at the same period with Mundinus, Robert, King of Naples, was a distinguished patron of medicine, and retained at his court two eminent physicians, Francis of Piedmont, and Matthew Sylvaticus, the latter of whom undertook to continue the supplement to the works of Mesue, which Petrus de Apono had commenced. Besides this completion of the unfinished labors of de Apono, Sylvaticus wrote a large volume around 1317, which he called *Pandectæ Medicinæ*, a kind of medical vocabulary, to facilitate the understanding of the translations of the Greek and Arabian writers. However, this work appears to have been, despite its formidable bulk, very poor in execution. He contributed more to the improvement of botany and to the accurate determination of the properties and uses of plants than any other writer of that dark period.

Medical science had until now neither made little if any progress in England, nor do we meet a single medical writer in the British islands before the time of Gilbertus Anglicanus. The reason of this backwardness of medical science among the English may be that there was no encouragement for proficiency in this department of knowledge at any of the colleges or universities, and the practice of medicine was wholly in the hands of the monks. Gilbertus, whose birthdate and birthplace are unknown, is the first practical writer on medicine, of whom the British islands can legitimately boast. He maintained a character both for erudition and talent, being deeply versed in philosophy and physic, and eminently successful in practice. Leland speaks highly of his skill and knowledge and particularly commends the rules he prescribes for the preservation of health and the perspicuity with which he explains the virtues of herbs. His writings partake largely in the universal defect of the age, being almost wholly compiled from the works of the Arabs, and on occasion exhibiting entire chapters taken directly from Rhazes. He also selects, but with acknowledgment, some of the best observations of Alexander. Gilbertus is said to have been the first English physician courageous enough to expose the absurd practices of the monks, who at that time dominated the practice of medicine. He contrasted their practice with that recommended by the ancients. Gilbertus's principal work was his *Compendium Medicinæ, tum Morborum universalium quam particularium*, which was corrected by Michael Capella and printed in Lyons in 1510, and a new edition appeared in Geneva in 1608 under the title *Laurea Anglicana, seu Compendium totius Medicinæ*. We do not know the time of his death.

John of Gaddesden

Not far distant from the time of Gilbertus lived the Englishman John of Gaddesden, author of the celebrated *Rosa Anglica*. He took the degree of Doctor of Medicine and flourished about 1320. This work was composed in the seventh year of his lectures, and it is likely that these lectures were read at his own Merton College, where a medical society appears to have been founded that encouraged the study of medicine more than all the other colleges of Oxford collectively. The cases that he introduces of a scholar and a noble bachelor to whom he attended with success appear to date the work somewhere between 1305 and 1317. He mentions Gordonius and is himself, as Leland testifies, spoken of by Matthew Sylvaticus in his *Pandects*. The *Rosa Anglica* embraces the whole practice of medicine, chiefly collected from the Arabian and Latin authors who wrote shortly before his time, to which John has added from his own practical knowledge. John of Gaddesden appears to have been the first English physician who was employed at the Royal Court, as before his time, the Royal physicians and druggists were all foreigners.

Glanville and Guido de Cauliaco

Bartholomew Glanville, the author of *De Proprietatibus Rerum*, is said by Pitt to have written a book on the practice of medicine. It is probable, however, that Pitt has confounded two individuals of the same name, for neither Leland nor Bayle mentions Glanville as the writer of any work of this description, nor do they allude to his having ever applied himself to the study of medicine. He treats of a variety of complaints in his seventh book, but this he copied principally from the writings of Constantine. Bartholomew, who wrote an abridgement of *Practice*, as the work was called, speaks of Glanville in such a manner as to make it evident that the same person could not be the author of both works. Bartholomew's *Breviary* is a voluminous work of fifteen long books. We meet in it the history of the propagation of elephantiasis that has been already noticed when speaking of the writings of Gilbertus. Both accounts are almost word for word the same, while nothing of the kind is to be found in Glanville. As for the rest of the work, the writer ingenuously admits that it has no claim whatever to originality, but is wholly compiled from such materials as he could find bearing upon his subject in the writings of the philosophers and the physicians.

Aided by the labors of the different writers of whom we have just spoken, as well as by a large share of practical experience, Guido de Cauliaco, or Guy de Chauliac (Figure 22), reduced the art of surgery to a regular system. Although he made few additions to the stock of knowledge previously acquired, he is deservedly considered a reviver of the languishing art of surgery. Guido was born in a small town of the Gavandon on the frontiers of Auvergne in the early part of the fourteenth century and studied medicine at Montpellier under Raimond de Molieres. He became a professor at Montpellier and practiced for many years in Leyden, after which he settled at Avignon, where he held the post of physician to Pope Clement VI and his successors. He only acknowledges having seen the sixth book of Paulus, but profited by its contents. He has also availed himself of the writings of others of his predecessors, especially among the Arabs, Albucasis in particular, which he has made use of with considerable judgment. He makes no mention of Celsus, whose works seem to have been but little known at this period. Among the books he speaks of having consulted, he mentions a translation of some of the books of Galen made by order of Robert, King of Sicily, by Nicolaus de Reggio. In the enumeration of authors whom he has consulted, Guido does not confine himself to giving a barren catalog of names and works, but annexes to each a critical examination of its merits.

Figure 22. Guy de Chauliac, d.c. 1370 (Image courtesy of Reynolds Historical Library, the University of Alabama at Birmingham).

Besides this catalog, he has left us a concise history of the state of surgery in his day, from which we learn that the surgeons of that period were divided into five sects, of which the first, following the lead of Roger, Roland, and the four masters, applied cataplasms indiscriminately to every ulcer and wound; the second, with Brunus and Theodoric, in similar cases employed wine only; the third, with William of Salicetus, and Lanfranc, treated wounds with emollient ointments and plasters; the fourth, forming the sect of Germans, who were mostly military surgeons, promiscuously employed oils, wool, potions, and charms; and the fifth, consisting of ignorant practitioners and silly women, relied upon all occasions to the saints, praised each other's writings perpetually, and followed each other in one undeviating track.

His *Magna Chirurgia*, which he completed in 1363, has been translated into most European languages. Guido was the first to point out the importance of making incisions over the eyebrows in the direction of the muscular fibers and not transversely. He also describes the lower end of the humerus and joint of the elbow more accurately than his predecessors. He revived the use of the trephine, and invented several instruments, of which he gives figures, among them a pair of forceps for taking up wounded arteries. The earliest account we have of the Cesarean operation is in Guido's *Chirurgia*, but he only speaks of it as resorted to after the death of the mother. It is likely, however, from the prevalence of the opinion that the operation had been performed before his time, although unnoticed by Hippocrates, Paulus, Celsus, or Albucasis, all of whom treat largely on the mode of assisting in cases of difficult parturition. On the subject of amputation, Guido agrees with Avicenna and Albucasis, that it is better to perform this operation on any extremity affected with gangrene than to sacrifice the life of the patient by omitting it. Also, Guido gives us an interesting account of the plague.

Ardern and Valescus de Taranta

John Ardern's birthplace or date have no authentic records, although he is likely to have been born in1319 or 1320, since we find him residing in full practice as a surgeon in Newark at the time of the plague in 1349. Such was his success in practice that his fame spread rapidly to London, to where he was induced to move in 1370. He has left a bulky volume of medicine and surgery, together with a number of manuscripts that have never been printed. Ardern may be regarded as the first reviver of surgery in England, and the numerous records of his work that he has left prove him to have been a man of great skill and extensive practice. He writes with an air of great simplicity. Although he adds quackery and superstition to his accounts, he was one of the best surgeons of his day. His productions contain many valuable observations. He offers a variety of pharmaceutical preparations, many of which were inventions of his own, and some of which retained their place in the pharmacopoeia. He invented a syringe for giving injections. He explains at length the utility of such injections not only in curing disease, but also in preventing disease, and we are led to conclude that the practice of giving injections was not common or well understood. He gives abundance of relevant cautions and says that the operator should not administer them rashly or negligently.

Valescus de Taranta, a native of Portugal and first physician to Charles VI of France, practiced medicine for 36 years at Montpellier and was one of the few writers of the age who dared to draw from the ample stores of his own experience. Although almost perfectly ignorant of Greek, he was an excellent Latin scholar, and around 1418 wrote a book entitled *Philonium*, which contains many observations on both medicine and surgery that merit attention. De Taranta was the first perhaps who proposed the treatment of cancer by an arsenic-containing preparation. He cautioned practitioners with respect to its use, as it was not totally safe. As de Taranta did not begin to write before 1418, after 36 years in practice, we can hardly suppose he commenced practice much before the age of 21 years, and we may without much fear of error mark 1361 as the year of his birth. His earliest work, *De Philonio*, was first printed in Venice in 1490 and again in Lyons in 1521 with a short essay subjoined. De Taranta appears to have been well acquainted with the doctrines of Galen and the Arabs.

Savonarola, Alexandra, and Colot

Such was the state of medicine as it regards practice at the close of the fourteenth and commencement of the fifteenth centuries. The properties and composition of the various mineral waters throughout the world began to attract a considerable portion of medical attention. The physicians of this period have handed down to us many observations of their nature and uses. Among the writers of this class Michael Savonarola stands eminent. In his addition to what J. de Dondis and Ugolino de Monte had written upon the subject, he gave an account of all the known mineral springs in Italy. He was a native of Padua and grandfather of the celebrated Brother Hieronymus. He had been physician to three Marquises of Ferrara and a knight of Jerusalem. He enjoyed a high reputation for skill and experience and wrote many books, especially one on fevers. It appears he commenced his work on mineral waters between 1440 and 1450. Sicily also produced about the beginning of the fifteenth century a physician of the name of Alexandro ab Alexandra, head physician to the island, on the duties of which office he composed a work titled *Constitutiones et Regulata Jurisdictionis regii Protomedicatus Sicilise elucidata*, which was written in 1429. In this century, Scarlet fever made its first appearance in Italy. In 1440 in Mainz, the art of printing was first discovered. From this discovery, medicine soon began to derive the most important advantages. Before this century, lithotomy was not practiced by any regularly educated surgeons. What kept general surgeons from interfering in this most legitimate and important part of surgery is not clear, unless it was out of deference to Hippocrates, who obliged his pupils to swear that they would not attempt the operation. Around 1460, Germain Colot, a surgeon in high favor with Louis XL, ingratiated himself with some of the itinerant practitioners, witnessed the operation, commenced a series of experiments upon dead bodies, and, when he had attained adequate proficiency in performing the operation, communicated his ideas to the physicians in attendance upon the court, through whose interest the King's permission was obtained for him to try his skill upon a condemned criminal. This man consented to undergo the operation on the condition of receiving his pardon if he should survive it. Colot performed it with the most complete success, and his patient became fully convalescent in 15 days.

Aurelianus

From the many barbarisms that disfigure his style, Le Clerc is inclined to assign a place in this century to Caelius Aurelianus, the last writer who appears to have adopted Methodist principles and translated the works of Soranus, one of the principal founders of the sect, adding a considerable number of observations, partly original and partly collected from various sources. This translation consists of eight books, three of which discuss acute conditions and five of which discuss chronic conditions. This work (Figure 23) has the merit of preserving fragments and explaining the doctrines of several ancient writers whose works would otherwise have been totally lost. He notices many conditions of which no mention is to be found in the writings of his predecessors, and many of his surgical observations are unique to himself. Aurelianus has given us a more copious description of the symptoms and progress of hydrophobia than any of his contemporaries or predecessors. Among other noteworthy works of this author are his mention of hydatids in some kinds of edema, his recommendation of oil injection for the cure of illness secondary to ascarides, and his remarks on the abuse of bitter and acrid substances in the cure of gout.

Figure 23. Sick people being treated with herbs, from “De Morbis Acutis et Chronicis” by Caelius Aurelianus, 1709 (Image courtesy of Wellcome Library, London).

Sylvius, Benedictus, and Barbaras

James du Bois, or Jacobus Sylvius, first saw the light at Amiens in France in 1478 and was educated by Tagaulitus. This man, having an elder brother at the College of Tournay in Paris who was a zealous promoter of literature, enjoyed the most favorable opportunity of mastering Greek and Latin. On selecting the profession of medicine, he applied himself assiduously to the works of the ancients, especially Hippocrates and Galen, of the latter of whom he became a zealous disciple and defender. However, he did not confine his studies to any single branch of the profession, but devoted his attention to all. He enriched the science of anatomy with a number of valuable discoveries. He was the first to detect those valves that he terms epiphyses, or membranous epiphyses, in the mouths of the azygos, jugular, brachial, and crural veins, as well as at the trunk of the vena cava. He also paid great attention to pharmacy and medical botany, for his improvement in which he took journeys to distant parts of the country in order to examine the plants in the places of their spontaneous growth.

On his return to Paris, he gave private lectures on the theory and practice of medicine from the works of Hippocrates and Galen. His success excited the envy of the Parisian physicians, who took advantage of his being without a degree, to compel him to discontinue lecturing, in consequence of which he was obliged to visit Montpellier for the purpose of graduating. He returned without a degree, but, succeeding in accommodating his differences with the Parisian professors, he finally obtained the degree of Bachelor of Medicine in 1531 and became lecturer at the College of Triquet four years later, while Fernelius lectured at the College of Cornouaille. In response probably to the anatomical and botanical demonstrations that Sylvius gave, but which Fernelius omitted, his class exhibited an overflow of pupils, while that of Fernelius was scantily attended. In 1550, he was appointed professor of medicine at the Royal College, a post he retained until his death in 1557 at the age of 77 years. As an anatomical writer, he had considerable claims to merit. Despite the scarcity of the opportunities he enjoyed for human dissection, the discoveries and improvements he made were important and were published in his works *Isagoge Anatomica* and *Observata in variis corporibus secandis*. He also wrote several works on pharmacy, in which he was fully proficient. He published a valuable edition of Mesue accompanied by his own translation and commentary. He was an established enemy of Vesalius, who, from having been his pupil, became his rival in anatomy. One of the grounds of his dissatisfaction with Vesalius arose from Vesalius correcting Galen. Somewhat later in this century, we find Alexander Benedictus, a teacher of anatomy and medicine at Padua. He was born in Verona and was a physician to the emperor Maximilian. He wrote *Alexandri Benedicti Physici Anatomia, sive Historia Corporis humani, ejusdem collectiones seu Aphorismi*, a work that has been often reprinted, and a paper on the plague.

Botany made some trifling improvement during this century. We find the study of plants revived by Hermolaus Barbarus, who first attempted to correct the numerous errors to be found in the works of Pliny and Dioscorides. However, no real progress was made towards establishing the science of Botany upon a rational or stable foundation before the following century. Barbarus died in 1493 at the age of 40 years.

The Fall of Constantinople

The capture of Constantinople by the Turks in 1453 exerted some influence over the state of medical knowledge. Multitudes of Christians, fleeing from that city, brought with them to Italy a large number of manuscripts by the Greek medical writers, and thus furnished the studious with an almost inexhaustible fund of materials for exposition and commentaries.

While thus engaged, a question naturally came about as to the degree or resemblance between the theories of the Arabian and the Greek writers, an inquiry that continued to occupy medical men for at least half a century, but resulted in little improvement of the science. During the greater part of this century, medical writers did little more than pirate the works of their predecessors and even of their contemporaries. However, surgery accomplished somewhat greater success.

Application of Chemistry to Medicine

During this century, as the rage for alchemy began to decline, the application of chemistry to medicine was conducive to the accomplishment of chemical discoveries. This was, in a great degree, owing to the exertions of Basil Valentine, a German monk who first discovered ammonia. Valentine also has the merit of first bringing antimony into notice as an important addition to the *materia medica*, and in his celebrated *Cumis triumphalis Antimoni* pointed out a number of valuable preparations of this metal, which, by their success in practice and the controversy they excited, gave a vigor to chemistry as never seen before. The art of printing also invigorated the field of chemistry. Isaac Hollandus, though retaining a deep tinge of the alchemical dye, directed the attention of his contemporaries and successors to improvements in the art of metallurgy and prepared the way for George Agricola, whose masterly *De lie metallica* made distinct this branch of chemistry from alchemy and laid the foundation of all the profitable discoveries that have followed. Such was the state of medicine in all its branches at the close of the fifteenth century.

Chapter 8

Medicine and Surgery in the Sixteenth Century

Almenar

Among the important writers of this century is George, or John, Almenar, a Spanish physician who published in 1512 a work on syphilis titled *Libellus de Morbo Gallico septem capitibus absolutus*, which was eagerly sought after. Almenar appears to have had an extensive experience with this disease, for which his chief remedies were warm bathing and mercurial frictions with an alternative syrup exhibited on the intermediate days.

Rhodion

Early in this century, Eucharius Rhodion, a physician of Frankfurt, published in German the first popular work on midwifery, the practical rules for which were all taken from Hippocrates and other ancient writers. However, from the impossibility of following some of them, it was evident he had never practiced the art. In place of removing them, a censure equally applicable to every direction up to that time was published, the whole of which tended more to mislead than inform, and to increase difficulty and danger. In 1532, Rhodion's work was translated into Latin, and soon after into French, English, and other languages. With all its defects, it continued as the only manual used in England by midwives before the appearance of Chainberlen's translation of Mariceau's work, containing an account of the first material improvement in the art, which was made around 1550 by Ambrose Pare.

Brissot

About this time, a remarkable controversy arose among the medical practitioners in Spain and Portugal pertaining to a deviation from the established method of bleeding in pleurisy, which was recommended and adopted by Peter Brissot, a French physician, who visited Portugal in 1518. Brissot was the son of a distinguished advocate who resided at Fontenai le

Comte in Poitou, where Brissot was born in 1478, and, having devoted a number of years to the study of philosophy and medicine, became a Doctor of Medicine in May 1514 at the University of Paris. He carefully examined the doctrines of Hippocrates and Galen, and finding, upon comparison with the writings of the Arabs, that these last had introduced many things unsanctioned by those great fathers of medicine, he proceeded to restore their doctrine and practice to its original purity, for which purpose he commenced with lecturing out of Galen, in place of Avicenna, Rhazes, and Mesue. He pointed out the error which prevailed among practitioners of bleeding in cases of pleuritis on the side opposite to the affected side, and not upon that which is affected. This practice he demonstrated, in a public disputation, to be repugnant to reason, and contradictory to the doctrines both of Galen and Hippocrates. He then traveled to the city of Ebora, Portugal, where he introduced his new method of bleeding in pleuritis, which appears from the beginning to have given great dissatisfaction and to have produced a long and intemperate epistle from Denys, who was physician to the king of Portugal. In answer to this letter, Brissot wrote a full justification of his innovation in practice, but which was prevented from appearing by the death of its author in 1522; however, three years later, it was printed in Paris by his friend Anthony Luceus. The dispute between Brissot and Denys was deemed important enough to warrant discussion before the University of Salamanca, where Denys's supporters succeeded in procuring a decree forbidding Brissot to practice until the question was decided. However, after deliberation and attentively hearing all the arguments on each side, the university came to a decision, Brissot's practice was correct and aligned with the principles of both Hippocrates and Galen.

Syphilis

The distinguished anatomist Jacobus Carpus, or Jacobus Berengarius, is supposed to have been the first to introduce the practice of friction into use, and is said to have accumulated much wealth by his practice. However, it seems that this treatment method for syphilis was at least partially adopted by Almenar some years before. Berengarius was born at Carpi in Modena, around the end of the fifteenth century. He was initiated into the practice of surgery by his father, who was an eminent surgeon. He went at a proper age to Bologna, and then to Padua, where he served as a professor of anatomy. In 1518, he returned to Bologna and was appointed as a professor of anatomy. He retired to the dominions of the Duke of Ferrara, where he died around 1527. The use of guaiacum for the cure of syphilis was first introduced by Gonzalo Fernandez, who, having contracted this disease at the Siege of Naples, was determined to seek the means of recovery in those regions from where the malady had originated, and embarked with this view for the West Indies. Fernandez, having experienced in America the salutary effects of guaiacum in counteracting the ravages of syphilis, availed himself of this knowledge on his return to Spain, and he made a fortune by providing this cure. He secured the monopoly of the remedy by carefully guarding the secret, until it was revealed to the world by Leonard Schmaus in 1518. The efficacy of the China root and sarsaparilla for the treatment of syphilis was made known by Aloysius Lobera, a Spaniard who was a physician of Emperor Charles V. The year 1530 was distinguished by the appearance of the celebrated poem on syphilis written by an eminent physician and accomplished scholar named Hieronymus Fracastorius. His poem was titled *Syphilidis, sive de Morbo Gallico*. Fracastorius was born in 1483 in Verona, where he practiced physic. He retired in his latter days to a country seat he had near Venice, where he died of apoplexy on August 6, 1553, at the age of 71 years. He was the author of several other medical works besides his celebrated poem on syphilis.

Linacre and the College of Physicians in London

The College of Physicians was founded in London by Thomas Linacre (Figure 24), an English physician of the time of Henry VIII, in the early sixteenth century. Linacre was born in Dover around 1460 and educated at Oxford, where he was elected a fellow of All Souls' College in 1484 at the early age of 24 years. Linacre traveled to Florence, where he experienced the most flattering reception from Lorenzo de Medici, who allowed Linacre to participate with his own sons in the instruction of their tutors, Demetrius Chalcondyla and Angelo Politiano, by which he made significant progress in learning Greek and Latin. He then studied philosophy and medicine, focusing particularly on mastering the doctrines of Aristotle and Galen, which no Englishman had ever attempted before. He translated several of Galen's works in the most masterly manner and elegant style. Soon after his return from Italy, he was selected by Henry VII, to superintend the health and education of the prince and princess at court.

Figure 24. Thomas Linacre, 1460-1524 (Image courtesy of Reynolds Historical Library, the University of Alabama at Birmingham).

Linacre gave medical lectures at both Oxford and Cambridge. His most important foundation was that of the College of Physicians. In those days, both the Bishop of London and the Dean of St. Paul's possessed the important privilege of examining and admitting medical candidates to practice within their respective jurisdictions. Other bishops claimed a similar power within their several dioceses. Linacre suggested the foundation of a college to be composed of men professionally capable of deciding upon the qualifications of the various candidates and the degree of their fitness to undertake the medical care of their patients. The College of Physicians was eventually established in 1523, and Linacre was the first president. The organization's meetings took place at his house. He was 64 years old when he died.

Paracelsus

Paracelsus (Figure 25) was born in Appenzel, Switzerland in 1493. He at first studied chemistry. While establishing to himself a high opinion of his own importance, he commenced a sort of rambling life, declaring that after the most careful examination of the principles of the medicine, he found them all erroneous, and so decided, after visiting the schools of France, Italy, and Germany, to travel in pursuit of truth, searching for it not only among the learned, but even among old women, mountebanks, barbers, and quacks of every description. After passing several years in this kind of manner, he succeeded in obtaining credit for an erudition that he never possessed, and for important acquisitions in science that he never made. It must in truth be admitted that he possessed the talent of imposing charlatanry in the most eminent degree.

By the bold manner in which he administered antimony, mercury, and opium, Paracelsus often succeeded in effecting cures that had baffled the more cautious practice of experienced practitioners. As his records reflect only successful cases and the more numerous failures were consigned to oblivion, he succeeded in creating a high opinion of his practical skill. His reputation at last attained to such a height that the magistrates of Basil were induced to engage him as a professor of medicine with a high salary in that city. During 1527 and 1528, he gave lectures, which at first attracted a numerous audience. However, his egotism, vanity, and excessive ignorance, which could not long be concealed, disgusted all the better informed among his pupils, and the number gradually diminished to zero attendees. As a proof of his ignorance and his arrogance, he commenced his very first lecture by publicly burning the works of Galen and Avicenna, impudently declaring that his cap contained more knowledge than all the physicians. Utterly unacquainted with either anatomy or physiology, and equally destitute of classical literature, he was unable to explain the defects of the ancients. As his theories were without system, so was his practice without judgment. In 1525, he relinquished his professorship in response to a disagreement respecting fees and abruptly left Basil to embrace once more a life of wandering, maintaining a kind of meteoric reputation by the occasional brilliancy of a few surprising cures. The merit has been awarded to Paracelsus, by some, of having been the first to employ mercury in the cure of Syphilis. However, this merit rests upon the same apocryphal grounds with his other pretensions. Paracelsus could not have exceeded his nineteenth year when Almenar wrote on this treatment for syphilis.

Figure 25. Paracelsus (from Guerini, Vincenzo. *A history of dentistry from the most ancient times until the end of the eighteenth century*. Philadelphia & New York, Lea & Febiger, 1909).

Bodenstein

Adam Bodenstein, who had been a pupil and one of the most strenuous defenders of Paracelsus, gave a fatal proof of his misplaced confidence in the doctrines and prescriptions of his master. Accordingly, the town of Basil was visited by a most destructive fever in 1577. Armed with an invincible Theriaca prepared according to a formula obtained from his master, Bodenstein was certain of vanquishing the disease. Instead of conquering the condition, he himself fell ill to it soon after his arrival in the city and died with the invaluable Theriaca in his possession, thus furnishing a striking example of the folly of blind credulity and overweening vanity.

Chapter 9

Anatomy and Physiology in the Sixteenth Century

Bonacciuoli's Enneas Muliebris

During the sixteenth century, anatomy and almost every other branch of knowledge experienced tremendous progress as a result of two events, the discovery of the art of printing and the Protestant Reformation. Greater progress was made in anatomy and physiology between the years 1500 and 1600 than in any period before. This period witnessed an appraisal by several anatomists of the circulation of the blood, which paved the path to Harvey's discovery.

The first notable anatomical work[1] in this century is *Enneas muliebris*, which was published in 1503. The first edition was dedicated to Lucrezia, the daughter of Pope Alexander VI. This work was the first to distinguish between the clitoris and the labia minora. *Enneas muliebris* was highly prized for its accurate anatomical descriptions of the fetus and the reproductive organs of both sexes. It is now, however, of little value.

Achillini

In 1516, Achillini first published in Venice *De humani corporis anatomia*, a work on the anatomy of the human body. In 1522, his annotations on the *Anatomy of Mundinus* made their appearance along with the *Fasciculus Medicinæ Johannis de Ketham*. He is thought by some to have discovered the malleus and the incus among the bones of the middle ear. Achillini was born in Bologna in 1472, and he was celebrated both as a physician and as a philosopher. He was appointed as a professor at a very early age and was promoted to the chair of philosophy in Padua in 1506. However, being compelled by the war between Venice and the league of Cambray to leave Padua, he returned to Bologna, where he was reappointed to the chair. He adopted and maintained the doctrines of Averrhoes and was the rival of Pomponacius. He died in Bologna and was interred in the Carmelite church of St. Martin the

[1] Enneas muliebris, sive de fœtûs formatione, &c. Fol. 1503. An 8vo. edition was published in 1537.

Great at the age of 40 years, and a complimentary epitaph was later penned to his memory by James Vitalis.

Berengarius

Berengarius developed the practice of salivation by means of mercurial friction in the treatment of syphilis. However, it is his important contribution to the study of human anatomy that has immortalized him under the name Carpus or Carpensis, a name derived from his birthplace of Carpï. Such was his passion for anatomical pursuits that he was accused of having dissected living human subjects for the purpose of observing peristalsis of the bowels. In 1521, Berengarius published commentaries upon *Anatomy of Mundinus*.[2] The following year he published *Isagoge breves, perlucide ac uberime, in Anatomiam humani corporis*[3] in Bologna. These two works contain all of his anatomical discoveries. He was the first to observe and discuss the appendix, to describe the anastomosis of the portal vein and vena cava within the liver, and to observe the caruncles in the kidneys.

Servetus

A few years after the anatomy of Berengarius, Michael Servetus,[4] a Spanish physician published two works, *De Trinitatis Erroribus* (On the Errors of the Trinity) in 1531 and *De Christianismi Restitutione* (The Restoration of Christianity) in 1533. These works may be regarded as promoting unorthodox views of Christianity, but established him as a philosophical inquirer and good anatomist. Nevertheless, the clergy viewed Servetus's writings as obnoxious. In *De Christianismi Restitutione*, we find one of the first insights given on the circulation of the blood prior to the work of Harvey. After observing that the vital spirit is composed of the most subtle parts of the blood and of the air that enters the lungs, and that this blood comes from the right ventricle of the heart, Servetus (Figure 26) writes, "But the communication, that is to say, the passage of the blood from the right to the left ventricle, does not take place across the middle septum, as persons have generally imagined; it depends upon a more singular structure. In the long winding of the lungs, this subtle blood is agitated, and prepared by the action of the viscus, and gains a yellow color. From the arterial vein (pulmonary artery) it passes into the venous artery (pulmonary vein), where it becomes mingled with the air that has entered the lungs, and loses its fuliginous excrements. Lastly, it enters the left ventricle, which attracts it in its diastole. Such is the

---

[2] Commentaria, cum amplissimis additionibus, supra Anatomiam Mundini, cum textu ejus in pristinum nitorem redacto. 4to. Bononiæ, 1521.

[3] Isagogæ breves, pellucidæ et uberrimæ, in Anatomiam humani corporis, ad suorum scholasticorum preces in lucem editæ. 4to. Bononiæ, 1522. 8vo. Coloniæ, 1529. 8vo. Argentorati, 1533, and 4to. Venetiis, 1535. A translation was published in London in 1664 by H. Jackson under the title of *A Description of the Body of Man, being a practical Anatomy*. Besides these works he had published a treatise *De cranii fractura*, 4to. Venetiis, 1518.

[4] For a more comprehensive review of the life and work of Michael Servetus, see the following article, Bosmia A, Watanabe K, Shoja MM, Loukas M, Tubbs RS. Michael Servetus (1511-1553), physician and heretic who described the pulmonary circulation. Int J Cardiol 2013, 167(2), 318-321. doi, 10.1016/j.ijcard.2012.06.046. Epub 2012 Jun 28.

preparation of the blood from which the vital spirit originates; this preparation, and this passage from the arterial vein, into the venous artery, are evidently proved by the size of the vessels, which would not be so large, or possess so many branches, or carry so great a volume of blood to the lungs, if it were merely designed for the nutriment of that viscus." He then continues that this vital spirit is sent from the left ventricle into the arteries of the body. It is clear that he was knowledgeable about the minor circulation of the lungs and had some idea about the chemical change that the blood undergoes from contacting the atmospheric air in the lungs.

Figure 26. Michael Servetus (Image courtesy of the National Library of Medicine).

Guinterius

Johannes Guinterius, whose real name was Winther, was a native of Andernach, a town in the district of Ubich in the circle of the lower Rhine. He published a valuable work on anatomy titled *Anatomicarum institutionum ex Galeni sententia*[5] in 1536. In this book, the term "pancreas" was first applied.

Guinterius was born in 1487 and sent at the age of 12 years to Utrecht to study. Falling on financial hardships, his studies were cut short, and he moved to Deventer, where he initially resorted to begging and working menial jobs. He overcame his difficulties and moved

[5] Anatomicarum institutionum ex Galeni sentential, per J. Guinterium Andernachum, Medicum. Libri quinque. Basileæ, 1536. 8vo. 1539, 4to.

to Marpurgh to study natural philosophy. He was appointed professor of Greek at the University of Louvain. He resigned from his professorship in Louvain to study medicine and moved to Paris in 1525. He obtained his bachelor's degree in 1528 and his medical degree in 1530. He was appointed as a physician to Francis I, and for many years practiced medicine in Paris, where he also gave private lectures on anatomy.

An offer from Christian III of Denmark persuaded Guinterius to move from Paris to Copenhagen, where he resided until religious upheaval in 1537 compelled him to move first to Metz and then to Strasburg. In Strasburg, he was appointed to a professorship at the university and became established in a lucrative practice and extended his visits occasionally to Germany and Italy. Letters of nobility were conferred upon him by the Emperor Ferdinand I in 1562. Two years before he died, he published *De Medicina veteri et nova*[6], in which he discussed the new and old systems of medical practice. Guinterius died at the age of 87 years on October 4, 1574.

The life of Guinterius provides an instructive lesson to the young, teaching them not to despair under the most discouraging circumstances or abandon hope under the severest calamities. There is no difficulty so great and no depression so low as to be beyond the counteracting influence of talents aided by industry, patience, prudence, and perseverance. It may not be in everyone's fortune to attain distinction, but it is possible for those who possess ordinary capacity to raise themselves and attain respectability.

Dryander

Contemporary with, but somewhat junior to Guinterius, was John Dryander, a native of Wetteran in Hesse and whose real name was Eichmens. He was educated in France and obtained his medical degree in Mayence. He settled in Marpurgh and taught anatomy from 1536 to the time of his death in 1560. He was the first to point out several distinctions between the cortical and medullary portions of the brain. He also observed the olfactory nerves but mistakenly called them the optic nerves. He published these observations in his book *Anatomise pars prior, in qua Membra ad Caput spectantia recensentur et delineantur*[7] in 1537. In 1541, he published a new edition of the *Anatomy of Mundinus*,[8] in which he frequently corrects Mundinus.

Vesalius

Vesalius was a native of Brussels and an eminent anatomist. His first work on the structure of the human body, *De humani corporis fabrica*,[9] debuted in Basil in 1543. He was the first to notice that the optic nerve was not attached directly along the axis of the eye, and that the round ligament of the thigh was not inserted into the middle of the head of the femur, but rather into its side. He also provided the first accurate description of the bones that

[6] De Medicina veteri et nova. 2 vols. Fol. Basileæ, 1571.
[7] Anatomiæ pars prior, in qua membra ad caput spectantia recensentur et delineantur. 4to. Marpurg, 1537.
[8] Anatomia Mundini ad vetustissimorum manuscriptorum codicum fidem collate. Marpurg, 1541.
[9] De humani corporis fabrica. Fol. Basileæ, 1543.

constitute the organs of hearing. Vesalius also wrote a book on the decoction of the China root,[10] a medication frequently used during that period to treat syphilis and other diseases.

Columbus

Realdus Columbus was a native of Cremona and a contemporary of Vesalius. Like Vesalius, he was distinguished for his work in anatomy. Concerning the circulation of the blood, he added to the imperfect sketch left by Servetus; he described the entrance of the blood from the vena cava and its subsequent transmission through the pulmonary circulation to the left ventricle and the aorta, and demonstrated that all of the blood, not only the vital spirit, passes through the lungs, thus elucidating this mechanism one step further than Servetus did. However, he still erroneously believed, as anatomists prior to him did, that the liver provided the blood supply to the stomach. Columbus was the first to describe accurately the caruncles in the vagina and discuss the parietal and visceral parts of the peritoneum and pleura. He claims to have discovered the tunica innominata of the eye, which was later, referred to as the tunica albuginea or innominata of Columbus, but Douglas believes Galen described this layer under the name "tunica sexta". He is credited with discovering the third bone in the organ of hearing. One of Columbus's students, Johannes Valverda, a Spanish anatomist, is said to have introduced the study of anatomy to his native country. Valverda published the anatomical plates of Vesalius, translated the descriptions into Spanish, and added four new figures that represented the direction and progress of the fibers composing the muscles of the anterior part of the body, a woman in a state of pregnancy, and the cutaneous veins of the anterior and posterior parts of the body.

Ingrassias

Another successful anatomist of this period was John Philip Ingrassias (Figure 27), a native of Sicily. He studied medicine at the University of Padua and received his medical degree in 1537. Following graduation, he accepted the chair of medicine and anatomy at Naples. He exhibited great practical skill in anatomy, especially with regards to the bones of the head and the organ of hearing. Ingrassias was the first to point out the stapes. Although others have claimed this discovery, Fallopius credits Ingrassias for it. Eloy was led to think from an inspection of his plates that he must have been acquainted with the muscle of the malleus, or tensor tympani, which Eustachius has been credited for discovering. The discovery of the seminal vessels is also ascribed to Ingrassias.

Eventually, Ingrassias returned to Sicily. In 1563, he was appointed as Proto-Medicus, or chief physician, of Sicily and its adjacent islands by Philip II of Spain. As chief physician, Ingrassias collectively held all the powers belonging to the Spanish College of Physicians, thereby enabling him to contribute to the improvement of medical science by rigorously examining the qualifications of prospective practitioners. His excellent services during the destructive plague that depopulated Palermo in 1575 bestowed upon him the honorable

[10] De modo propinandi Radicis Chinæ Decoctum. Fol. Basileæ, 1546.

appellation of the "Sicilian Hippocrates". Ingrassias published 10 works during his life.[11] His principal work was published posthumously in Messina in 1603 under the supervision of his nephew, Nicholas Ingrassias. This posthumous work, *In Galeni librum de ossibus doctissima, et expertissima commentaria*, contained a commentary on Galen's book on the bones. Ingrassias died in 1580 at the age of 80 years.

Figure 27. John Philip Ingrassias (from Guerini, Vincenzo. *A history of dentistry from the most ancient times until the end of the eighteenth century*. Philadelphia & New York, Lea & Febiger, 1909).

[11] The following is a list of the several works published by Ingrassias during his lifetime, Iatropologia; Liber quo multa adversus Barbars Medicos disputantur, 8vo. Venet. 1544 – 1558. Scholia in Iatropologiam. Neap. 8vo. 1549. De tumoribus præter naturam, Fol. Neap. 1553. This, in fact, is a commentary on some of the works of Avicenna. Raggionamento fatto sopra l'infermità epidemica dell'anno 1558. 4to. Palermo 1560. To this work was subjoined the following, Trattato di due mostri nati in Palermo in diversi tempi. Constitutiones et Capitula, nec non Jurisdictiones Regii Proto-Medicatûs officii, cum Pandectis ejusdem reformatis, 4to. Panormo, 1564 – 1567. In this he gives an interesting account of the nature and duties of the office to which he had himself been appointed the preceding year. Quæstio de purgation per medicamenta, atque obiter etiam de sanguinis missione, an sexta die possit fieri, 4to. Venet. 1548. Galeni ars medica, fol. Ven. 1573. De frigido potu post medicamentum purgans Epistola, 4to. Ven. 1575. Informazione del pestifero e contagioso morbo, il quale affligge e habe afflitto la città di Palermo e molti alter città e terre del regno di Sicilia, nell' anno 1575 e 1576, 4to. Palermo 1576 – giving an account of the dreadful visitation in which he distinguished himself so much for zeal and ability. A Latin translation of this was published in 1583, at Nuremberg, by Joachim Camerarius under the name of Methodus curandi pestiferum contagium.

Fallopius

In 1548, Pisa received one of the most distinguished anatomists and physicians of this century, Gabriel Fallopius (Figure 28), as a professor of anatomy. Fallopius was believed to be born at or near Modena in 1523. After studying anatomy under Brassavola, he travelled to other countries to improve his knowledge base. He acquired an astonishing amount of knowledge at a very early age and mastered not only medicine and anatomy, but also botany, astrology, and chemistry. He was only 25-years-old when he obtained a professorship at Pisa, yet he moved to Padua around 1551. He has been described as being "most methodical in teaching, most successful in healing, and most expeditious in dissecting." His endeavors contributed greatly to the elucidation of anatomy, although he has claimed and obtained credit for discoveries made long before his time. For example, the uterine tubes named after him, the Fallopian tubes, were described by Rufus of Ephesus and Herophilus, and his claim to the discovery of the pyramidalis muscle is unfounded, as it was observed by Galen and Jacobus Sylvius. Fallopius's *Observationes anatomicæ*,[12] which was published in 1561, is one of the best works of that century. Fallopius died at the age of 40 years in 1563.

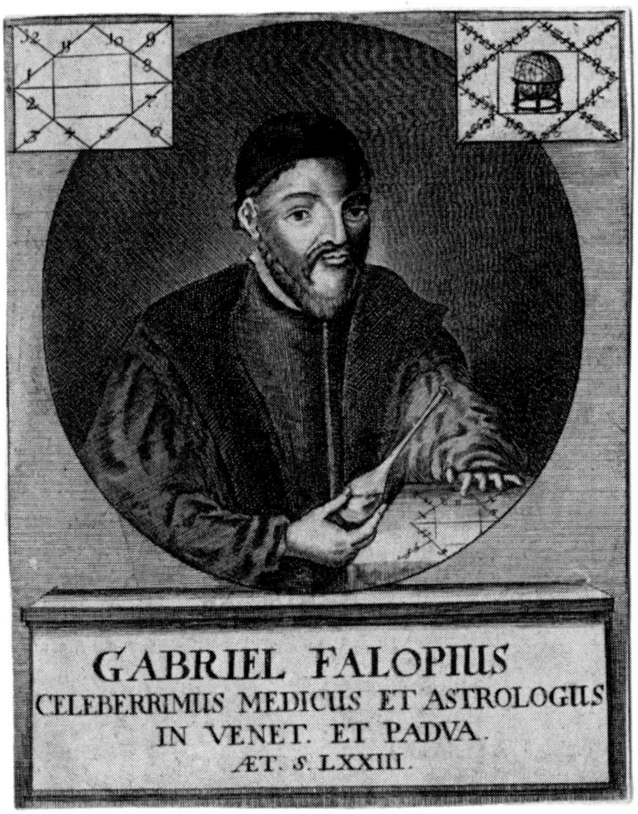

Figure 28. Gabriel Fallopius, 1490-1563 (Image courtesy of Reynolds Historical Library, the University of Alabama at Birmingham).

[12] Observationes anatomicæ in libros quinque digest. Venetiis, 1561.

Eustachius

Bartholomew Eustachius (Figure 29) is known for his work *Opuscula Anatomica*,[13] which was published in Venice in 1563. He was born in the small Italian village of San Severino probably around 1520. He studied anatomy in Rome with such success that he was appointed to the chair of anatomy at that college. He died there in 1574. Most of his works are lost. To him we are indebted for the discovery of the adrenal glands. Eustachius was critical of Vesalius in some instances. He blames Vesalius for not acknowledging that he had examined the kidney of a dog in place of that of a human and had not noticed the difference between the two. He claims to have discovered the stapes, but Fallopius attributes this discovery to Ingrassias. He was the first to describe accurately the thoracic duct, which he says resembles a white vein in horses and opens with a semicircular orifice into the internal jugular vein. He discovered the pharyngotympanic tube, which is called the Eustachian tube in his honor, and the valve at the orifice of the coronary vein of the heart. He claims to have discovered the valve in the vena cava near the right auricle of the heart. However, this structure appears to have been noted by Sylvius before him.

Figure 29. Bartholomew Eustachius (from Guerini, Vincenzo. *A history of dentistry from the most ancient times until the end of the eighteenth century*. Philadelphia & New York, Lea & Febiger, 1909).

[13] Opuscula Anatomica, nempe de renum structura, officio, et administratione, de auditûs organo, ossium examen, de motu capitis, de vena quæ azygos dicitur, et de alia quæ in flexu brachii commune profundam producit, de dentibus. Venetiis, 1563.

Arantius

Julius Caesar Arantius was a native of Bologna and born around 1530. Vesalius and his uncle Bartholomew Maggius educated him. His first work, *De humano fœtu opusculum*, was published in Rome in 1564. In this text, he corrected many errors into which preceding anatomists had fallen regarding the uterus and the origin of its vessels. In the first chapter of the fourth edition of this work, which was published in Venice in 1587, he describes the substance of the uterus as resembling a sponge, divisible into lamina, and perforated with minute pores. He describes the vessels of the uterus in the third chapter. In the fourth chapter, he describes the foramen ovale and ductus arteriosus in the fetus and denies the existence of the urachus or allantois in humans.

In 1579, Arantius published a commentary on Hippocrates' book on wounds in the head. In his work *De tumoribus præter naturam*, which was published in 1581, Arantius described a pair of forceps that he had invented for extracting polyps from the nose. In this work, by having demonstrated that there was no communication after birth between the right ventricle and the left ventricle and that the blood was carried through the lungs by the pulmonary artery, he went one step further than most of his predecessors towards elucidating the blood's circulation. However, Servetus and Columbus had already demonstrated this fact before him. Until his death in 1589, Arantius held the chair of medicine, anatomy, and surgery at the University of Bologna, from which he had received his medical degree.

Coiter

Volcherius Coiter (Figure 30) was a native of Groningen. He was a pupil of both Fallopius and Eustachius. After completing his studies at Padua, Rome, Bologna, and Montpellier, he became a physician for the French Army in order to increase his chances of obtaining subjects for dissection. He made numerous observations on the changes in the viscera caused by different diseases. He contributed largely to the improvement of anatomy and surgery, among the former of which were his observations on the brain, whose motion he found to originate in that of the arteries. He contended that the brain was not always essential for life, as some animals survived its removal. Coiter was the first to describe the corpora lutea in the ovaries along with the order in which the parts of a chick develop in the egg. With respect to the heart, he observed that the contraction of the auricle follows, and does not precede, that of the ventricle. His descriptions of the frontal sinuses and the organs of hearing are more detailed than those of any other author. He discovered, or at least appears to have been the first to describe, that the two corrugators of the eyelids are distinct muscles along with two similar muscles of the lips. He prepared skeletons from various age groups to explain the origin of the bones and demonstrate the difference between the bones of an infant and those of an adult.

Figure 30. Volcherius Coiter (from Guerini, Vincenzo. *A history of dentistry from the most ancient times until the end of the eighteenth century*. Philadelphia & New York, Lea & Febiger, 1909).

In 1566, he published *De cartilaginibus, tabula quinque.*[14] After leaving the French Army, he settled in Nuremberg, where he published his principal work on anatomy,[15] in which he included a short account of the history and progress of anatomical knowledge and the order in which it should be studied. In this work, he provided a complete set of plates of the skeleton of a fetus and explained his observations on the growth of the bones. In 1575, Coiter published his work on comparative anatomy.[16] Among his surgical observations, Coiter noted that injuries of the brain in which the dura mater remains unruptured pose

[14] De cartilaginibus, Tabulæ quinque, Fol. Bononiæ, 1566.

[15] Externarum et internarum principalium corporis humani partium tabulæ atque anatomicæ exercitationes, observationesque variæ, novis et artificiosissimis figuris illustratæ, Fol. Norimbergi, 1573.

[16] Diversorum animalium sceletorium explications, cum lectionibus Fallopii de partibus similaribus. Fol. Norimb. 1575.

greater danger than injuries in which it has ruptured. Therefore, he would open the dura mater to let out the extravasated matter. He died in 1576 at the age of 42 years.

Varolius

Constantius Varolius was a native of Bologna. He was known for his skill in anatomy, and he discovered a part of the brain that has been named the pons Varolii or Varolius's bridge in his honor. He was the first to divide the brain into three portions by adding the medulla oblongata. He is also credited with discovering the glands of the choroid plexus and the valve of the colon.

Caesalpinus

Several authors attempted to describe the circulation of the blood prior to Harvey's discovery. Columbus improved upon the demonstrations of Servetus, and Arantius upon those of Columbus. The anatomist Andreas Caesalpinus (Figure 31) came closer still to the truth. He was distinguished for his skill in anatomy and for his work in botany.

Figure 31. Andreas Caesalpinus (Image courtesy of the National Library of Medicine).

He wrote *Quæstionum Peripateticarum*[17] in 1571 in opposition to the doctrines of Galen. In this work, he reasons that the pulsation of the arteries arises from the blood in the heart. However, he is still more explicit in the fourth chapter of the fifth book of the aforementioned work when he describes the minor circulation through the lungs, which Servetus first distinctly demonstrated. He clearly describes the manner in which the blood circulates from one ventricle of the heart to the other and shows that he understood the nature and function of the valves. He approached so close to discovering the return of the blood from the arteries through the veins that he even noticed the fact of the veins swelling below a ligature. However, he stopped short here, and he mistakenly believed that the blood moved backwards and forewards in the same vessels and returned to the heart during sleep.

Piccoluomini and Albertus

Archangelo Piccoluomini was a native of Ferrara and a citizen of Rome. He published a work on anatomy titled *Anatomia, sive de resolution corporis humani*[18] in 1573. It mentions the three valves at the entrance of the cecum that prevent the retrograde motion of fecal matter. Piccoluomini was the first to distinguish the grey matter from the white medullary matter of the brain, to delineate the anastomosis of the portal vein and vena cava within the liver,[19] and to describe that membrane of the fat later termed the "adipose membrane" by Riolan. Piccoluomini left a volume of anatomical lectures[20] and commentaries on Galen,[21] which was published in 1586 and 1556, respectively.

Salomon Albertus discovered the valve of the colon, which acts to prevent the regurgitation of fecal matter. Albertus was a professor of physic at Wirtemberg and published a textbook on anatomy[22] in 1583. In this book, he speaks of having first observed the aforementioned valve in a beaver and in a man. Furthermore, Albertus distinguished the small intestine from the large intestine.

Fabricius

Both Arantius and Caesalpinus knew that the valves of the veins exist. However, their structure and function were not clearly demonstrated before the year 1574, a little more than half a century before the first publication of Harvey's discovery, when Geronimo Fabricio, who is better known as Hieronimus Fabricius ab Aquapendente (Figure 32), gave the first satisfactory description of their structure and exhibited their first precise delineation in his engravings. Fabricius was born in 1537 to impoverished parents in the little town of

---

[17] Quæstionum Peripateticarum, libri quinque, 4to. Venetiis, 1571.

[18] Anatomiæ, sive de resoluti, one corporis humani libri quatuor. 8vo. Patavæ. 1573. A second edition was published, also in 8vo. at Frankfort in 1591.

[19] Hamilton writes that Berengarius provided the first description of the anastomosis of the portal vein and the vena cava.

[20] Prælectiones Anatomicæ. Fol. Romæ, 1586.

[21] Commentarii in Librum Galeni de humoribus. 8vo. Parisiis, 1556.

[22] Historia plerarumque corporis humani partium, in usum tyronum, 8vo. Wittebergiæ 1583, 1602, 1630.

Aquapendente in the district of Orvieto in Tuscany. His parents sent him to Padua, where he studied anatomy and surgery under the celebrated Fallopius. Upon Fallopius's death in 1563, Fabricius was appointed successor to the chair of anatomy at Padua. He held this position for nearly half a century. Fabricius attracted many pupils, one of whom was Harvey (Figure 33), who visited Padua in 1598. Fabricius was held in high esteem, and by order of the Republic of Venice, his name was inscribed upon a spacious anatomical amphitheatre erected for his accommodation. He was also made a knight of the Order of St. Mark by the Republic of Venice.

Figure 32. Hieronimus Fabricius ab Aquapendente (Image courtesy of the National Library of Medicine).

Fabricius directed his attention mainly towards anatomy and surgery, on which his publications were both numerous and important.[23] His *Opera anatomica*, which was published in 1603, contains an essay on the language of brutes well worth the attention of naturalists. His thorough knowledge of anatomy enabled him to improve the practice of surgery, thereby earning him the title "Father of Modern Surgery". His *Opera chirurgica*, which was published in 1618, was highly regarded. Fabricius died at the age of 82 years in 1619.

[23] See the original text for a list of Fabricius's works on anatomy and surgery.

Figure 33. William Harvey, 1578-1657 (Image courtesy of Reynolds Historical Library, the University of Alabama at Birmingham).

Chapter 10

Botany, Chemistry, and Pharmacy in the Sixteenth Century

Anguillara

We have the testimony of Homer that knowledge of the medicinal properties of herbs was esteemed as a high honor to possess. However, with the exception of the casual introduction of the subject by Homer, we seek in vain for information respecting the ancients' knowledge of botany from anyone besides the writers on medicine, even the best of whom provide descriptions that are so vague that it is almost impossible in the present day to determine which plants they spoke of. During the long and dismal eclipse that followed the overthrow of the Western Empire, knowledge of every description found an asylum at the courts of the caliphs, where medicine was successfully cultivated by the Arabian physicians. With the exception of adding a few vegetable simples to the scanty *materia medica* of the age, they made no farther progress in botanical research than their predecessors. When literature began to revive in the West, men naturally regarded the writings of the ancients as the depositories of nearly all the existing knowledge of the time, and therefore naturally directed their attention exclusively to the examination and elucidation of these writings, but examining the writings of ancient writers such as Pliny, Dioscorides, and Theophrastus for detailed information on botany was not fruitful. It was not until the end of the fifteenth and the start of the sixteenth centuries that the science of botany could be truly said to assume a consistent form.

The first person whom we find distinguishing himself in the field of botany during the sixteenth century is Louis Anguillara, who appears to have been the first Italian who traversed Greece, Cyprus, Candia, Switzerland, and other countries, for the express purpose of studying the properties of plants. He succeeded at this task so far as to have been able, as Haller informs us, to correct the works of Dioscorides and Matthiolus. Upon returning to Italy, he was appointed curator of the botanical garden in Padua, and he held this post until his death in 1550, leaving behind a book written in Italian on the knowledge of simples and published in Venice by Murinelli in 1561. A Latin translation by Caspar Bauhin was printed in at Basil in 1593.

Brunsfels and Fuchsius

Amongst the restorers of botany who preceded the illustrious Gesner was Otho Brunsfels, who in 1531 published the first two volumes of his great work on botany,[1] but which was not completed before 1536, when the third and last volume made its appearance. In 1532, he published a commonplace medical book[2] in which he addressed the use of drugs. In 1534, he published his *Onomasticon Medicince.*[3]

In 1542, Leonard Fuchsius, a native of Wembling, Bavaria, published his *History of Plants*, in which he chiefly copied Dioscorides and added a few remarks of his own. In this work, he falls into the prevailing error of the age, namely expecting to find the plants of those fairer regions in which the ancients wrote beneath the frowning skies of his own climate, which was less favorable for their growth. Like Brunsfels, Fuchsius had embraced the tenets of the Reformation, but upon finding his situation at Ingolstadt uncomfortable on a religious account, he resigned as professor of medicine and moved in 1528, at the invitation of George Margrave of Bareuth, to Onolezbach, where he was particularly successful in practice and adept at treating the sweating sickness, which began to prevail in Germany in 1529. He resided there for five years, during which time he published a compendium or introduction to the practice of physic, a Latin translation of the sixth book of Hippocrates with a commentary, and an apology against Triverius, a physician of Louvain, recommending bleeding of the right side in pleuritis and other visceral inflammations. In 1535, he accepted an invitation from Ulric, Duke of Wirtemberg, to settle at his University in Tubingen, where the Duke wished to form an asylum for all who chose to separate from the Roman Catholic Church. After living there for 35 years, he died on May 10, 1566, at the age of 65 years.

Gesner

Before the days of Linnaeus, botany was principally indebted to Conrad Gesner for reducing it to a systematic form and establishing its principles. Gesner was born in Zurich in 1516. He was the son of a hide worker who died during the Swiss Wars, thereby leaving him in such poverty that he was obliged to go to Strasburg to work as a servant. His master, observing his fondness for academics, allowed him to study whenever he did not need his services, by which means he made such progress that after accumulating some money, he went to Paris, where he applied to study philosophy and medicine, but he was obliged to return home and become a teacher because he lacked resources. He was so successful that he was able to resume his studies at Montpellier, and obtain the degree of Doctor of Medicine at Basil in 1540. He settled in Zurich, where he was appointed professor of philosophy and happily taught his fellow citizens for 24 years.

At an early age his maternal uncle Friccius fostered in him a love for plants. He founded and maintained a botanical garden, in which he cultivated all the rare or valuable plants that he could collect. He first discovered the necessity of dividing plants into classes, genera, and

[1] Herbarum vivæ icons ad Naturæ imitationem summa cum diligentia et artificio efficiatæ, cum effectibus earundem. Fol. Basil. 1-2. 1531-3. 1536.

[2] Theses seu communes loci totius Medicinæ, etiam de usu pharmacorum. Argentinæ, 1534.

[3] Onomasticon Medicinæ, nomina continens omnium stirpium, &c. Fol. Argentoreti, 1534.

species, and the distinguishing characters of each he took from the flower and the fruit. He also carefully examined the medicinal properties of plants and often endangered his own health in designing experiments for the benefit of others. It was at one time reported and believed that he had killed himself with the root of Doronicum, but he recovered. He eventually contracted the plague while attending to the sick and died on December 13, 1565. His work *De raris et admirandis herbis, quæ, sive quod nocte luceant, sive alias ob causas Lunariæ nominantur*, which was printed in Zurich in 1555, is a curious work with woodcuts and provides a description of Mount Pilate in the northern Alps, which he explored that year, together with notices of some Alpine plants. His *Historia plantarum et vires* was published in Basil in 1541. He also wrote a large number of works on various parts of natural history and botany. His remarks on the scientific arrangement of plants, which constitute his distinguishing merit, are chiefly dispersed among his letters, which were published posthumously.

Clusius and Turner

Charles Clusius, or more correctly de l'Ecluse, was a native of Arras in the French Netherlands. He was born on February 19, 1526. He attempted to develop a scientific arrangement of plants according to their size, habit, and sensible qualities. In 1576, he published an account of the rare plants that he observed in Spain,[4] embellished with 220 wood engravings of plants.

In this work, he points out fructification as being of primary importance for determining the genera of plants, as Gesner had explained. For the Alpine regions, he furnished a similar work in his history of rare plants[5] observed in Pannonia, Austria, and the neighboring districts, embellished with 350 wood engravings. In 1601, both works were reprinted in Antwerp[6] with considerable additions, such as a treatise on fungi, some of Clusius's correspondence, and Ponas's account of Mount Baldus. In 1593, at the age of 68 years, he was appointed to the botanical chair at Leyden, which he retained until his death. His third publication was an illustrated set of 10 books on exotic plants.[7] This work is chiefly founded upon the observations of Garcias ab Orta, Acosta, Monardez, and Bellon, and contains illustrations and an appendix on rare plants by Clusius himself. Clusius died at the age of 84 years on April 4, 1609.

Much praise is due to Dr. William Turner, an English contributor to botany who published the work *A New Herbal*[8] in 1551, for having given names to many English plants and for his diligence in examining and discriminating among the different species.

[4] Rariorum aliquot stirpium per Hispanias observatarum historia, 8vo. Antwerp, 1576.

[5] Rariorum aliquot stirpium per Pannoniam, Austriam, et vicinas quasdam Provincias observatarum historia, 8vo. Antwerp, 1583.

[6] Rariorum plantarum historia, fol. Antw. 1601.

[7] Exoticorum libri decem, fol. Antw. 1605.

[8] A New Herbal, two parts, fol. Lond. 1551.

Camerarius

Joachim Camerarius, a learned and eminent physician of Nuremberg, was born in 1534. He established a medical college at his birthplace and formed an extensive garden for the cultivation of botany. With a view to disseminate botanical knowledge, he published a work on botany and rural affairs[9] in 1677. Having purchased, methodized, corrected, and enlarged the collections of Gesner, and Wolfe, in 1586 he gave them to the world in conjunction with the works of Matthiolus.[10] In 1588, he published a work[11] in which he gave brief descriptions of a large number of plants illustrated by many new figures. He also published a medical tract on the best means of escaping the contagion of the plague.[12] He died on October 11, 1598.

Caesalpinus

Andrew Caesalpin, or Caesalpinus, of Arezzo in Tuscany was born in 1519. He was the first who began to execute the grand conceptions of Gesner and who conceived of the project to classify all the plants then known based on the various forms of the fruit, with classification being modified occasionally by the consideration of other external parts. He is the first writer who clearly distinguishes the sexes of plants. Caesalpinus derived his love of botany from Luke Ghinez, director of the botanical garden at Pisa, under whom he was educated. Having obtaining his medical degree at Pisa, he obtained the professorship of anatomy and medicine there, which he retained for many years. Pope Clement VI made him his first physician.

In his sixteen books on plants,[13] he compares seeds to the eggs of animals, the see serving to protect and nourish the embryo, or germ, until it had taken root. The value of this work is greatly lessened by the lack of figures and proper names. In 1603, he published an appendix[14] to this work, which constitutes the whole of his botanical writings. In his publication on the properties of medicines,[15] he says that he considers bleeding useful only at the commencement of fevers and recommends clearing out the stomach and bowels. Caesalpinus died in Rome on February 23, 1603, at the age of 84 years.

Bauderon and Baccius

Pharmacy does not appear to have progressed with any great success during the sixteenth century. However, among the pharmacists of the day, Brice Bauderon, a native of Charolles, merits notice for his acquaintance with this branch of medical knowledge. In 1588, he

---

9 Opercula de re rustica, quibus, præter alia, catalogus rei botanicæ et rustic, veterum et recentiorum, insertus est, 4to. 1577.

10 De plantis utilissimis, Petri Andreæ Matthioli novis iconibus et descriptionibus plurimis diligenter aucta, 4to.

11 Hortus medicus et philosophicus, in qua plurimarum stirpium breves descriptions, novæ icons non paucæ, continentur, 4to.

12 De recta et necessaria ratione preservandi a pestis contagion. 1583.

13 De plantis libri sedecim, 4to. Florentiæ, 1583.

14 Appendix ad libros de plantis, 4to. Romæ, 1603.

15 De facultatibus medicamentorum, libri duo, 4to. Venetiis, 1593.

published a pharmacopoeia founded upon the pharmacopoeia Lyonnensis with the observations of Catalanus on distilled waters. For a long period this book was the standard work on pharmacy throughout France. It was translated into Latin by Philemon Holland and published in London in 1639, since which there have been various editions both in French and Latin. Bauderon practiced physic at Macon, where he wrote a work on the practice of medicine,[16] which was printed in Paris in 1620. Among other contributors to the pharmaceutical knowledge of this period is Andrew Baccius, a native of Ancona and physician to Cardinal Ascanio Colonna, who afterwards became Pope Sixtus V. He wrote *De thermis, lacubus, fluminibus, et balneis totius orbis*, which was published in Venice in 1571; a paper on poisons and their antidotes, which was published in Rome in 1586; and a small account of the medicinal uses of precious stones.

John Bauhin

Among the writers on botany at this period, none distinguished themselves more successfully than the family of Bauhins. John Bauhin was born in Lyons in 1541. After the completion of his preliminary education, he was sent at the age of 20 years to accompany the illustrious Gesner in his botanical excursions through Europe. In the course of these, Bauhin collected many plants, which constituted the basis of his great work,[17] which was published posthumously. On his return from these excursions, he settled in Basil, where he was elected professor of medicine in 1566. However, he moved soon after to Yverdun, which he left for Montbelliard on the invitation of the Duke of Wirtemberg to become his chief physician. Here he passed the remaining 40 years of his life. In 1591, his book on plants named after the gods or saints[18] was published by his brother Gaspard in Basil. The elementary version[19] of his great work was published in Yverdun in 1619 and contains the rudiments of a natural classification of plants. However, it is for his posthumously published work *Historia plantarurm nova et absolutissima, cum auctorum consensu et dissensu circa eas* that he earned the deserved rank he holds among the improvers of botany. John Bauhin died in Montbelliard in the year 1613 at the age of 72 years.

Gaspard Bauhin

John Bauhin's younger brother, Gaspard, completed his preliminary studies under Fabricius and other distinguished anatomists in Padua, Montpellier, and Paris, and collected many plants that had eluded his brother and Gesner. Gaspard Bauhin returned from his travels to his native city of Basil in 1580. There he earned his medical degree and was appointed two years later as a professor of Greek. In 1588, he became a professor of botany and anatomy, and afterwards of medicine. He was later appointed chief physician to the city of Basil, dean

[16] Praxis in duos tractatus distincta, in priore agitur de febribus essentialibus, tam simplicibus, quam compositis, confuses, erraticis, malignis, ac pestiferis, et symptomaticis in genere et specie, curandis, in posterior de symptomatibus et morbis internis a capite ad pedes usque, 4t. Paris, 1620.

[17] Historia plantarum universalis, 3 vol. Fol. Ebroduni. i and ii. 1650. iii. 1651.

[18] De plantis a divis sanctisve nomen habentibus. Basileæ, 1591.

[19] Historiæ Plantarum generalis Prodromus. Ebroduni, 1619.

of the faculty of medicine, and rector of the university. He had previously published a Latin translation of Rousset's book in favor of the Cesarean operation, adding to it a multitude of cases and observations, including the history of a woman who had successfully undergone the operation, and a description of the valve of the colon. In 1596, he published his *Phytopinax*,[20] in which he gives general names of plants described from the specimens preserved in his Herbaria. Moreover, his edition of the works of Matthiolus[21] published in 1598 lists, as Haller observes, a considerable number of novel plants. His catalog[22] of indigenous plants growing in Basil is the largest catalog of plants growing spontaneously in a single district. Besides his numerous botanical works, he published one on anatomy that is illustrated by a volume of plates, the latter of which are chiefly copied from Vesalius, while the former contains an account of several of his anatomical discoveries, such as the contraction of the right side of the colon.

Colonna

In 1592, another attempt was made to determine the plants mentioned by the ancients by a comparison of living specimens with their vague and unsatisfactory descriptions in a work[23] published by Fabius Columna, or Colonna. He was born in Naples in 1567 to the noble House of Colonna and, as Boerhaave tells us, studied the works of Dioscorides in the hope of discovering a cure for epilepsy, with which he was troubled, but for which he is said to have found the application of a caustic to the thigh by Severinus, in 1630, a much more effectual remedy than all the simples described by Dioscorides. In this work, which he wrote at the age of 24 years, he has most accurately determined the character of wild valerian, which is, as Haller says, the plant whose root Dioscorides recommended in epilepsy, and which Colonna took according to his directions for a long time. As a scholar of botany, Colonna is entitled to the praise bestowed upon him by Haller, especially for illumination of the writings of Dioscorides that his works achieved and the number of plants that he has described. In 1606, he prepared another botanical work[24] descriptive of the rarer plants of Italy, with a farther inquiry into those that correspond with the descriptions of the ancients. A second part soon followed. He now began to delineate the flowers, fruits, and seeds of plants, and use their similarity as the basis of a system of classification. In his annotations on the work of Franciscus Hernandez,[25] he makes still greater advances towards that admirable system of classification that Linnaeus adopted. The works of Colonna were republished in 1744 by James Plaucus and accompanied with observations by the Lyncei, a society of naturalists of which he had been a member, and a biographical sketch of the author. Colonna died in 1651 at the age of 83 years.

[20] Phytopinax, seu Enumeratio plantarum ab herbariis descriptarum, 4to. Basil, 1596.

[21] Matthioli Opera quæ extant Omnia. Fol. Francf. 1598.

[22] Catalogus Plantarum circa Basileam sponte crescentium, 8vo. Bas. 1622.

[23] Plantarum aliquot historia, in qua describuntur Plantæ rariores antiquorum delineatienibus respondents, 4to. Neapol. 1592.

[24] Minus cognitarum rariorum que nostro cœlo orientium stirpium, ecphrasis, qua non paucæ ab antiquioribus descriptæ disquiruntur et declarantur, 4to. Romæ, 1616.

[25] Adnotationes et Additiones ad Opus Francisci Hernandez, et Nardi Antonii Recchi. 1627.

Gerarde

John Gerarde authored *The Herbal or General History of Plants*, which was published in London in 1597. The basis of this work was the *Stirpium Historiæ sex Pemptades* of Dodonæus. Gerarde was born in Nantwich, Cheshire, in 1545. According to Grainger, he was chief gardener for many years to the celebrated Lord Burleigh, minister to Queen Elizabeth, who possessed the first collection of exotic plants in England, many of which were introduced by Gerarde. Gerarde is well known to have been a surgeon of considerable eminence, talents, and erudition. He resided in Holborn, where he had an extensive botanical garden on which he published a catalog in 1596.[26] Gerarde significantly contributed to the improvement of botany in England not only by introducing numerous exotic plants, but also by promoting the discovery of indigenous plants not previously known. From a historical preface found in 1607 within a second edition of Gerarde's *Herbal* and published by Dr. Thomas Johnson, we learn that Gerarde died in 1607.

Alpinus

Prosper Alpinus was the son of a physician, and junior to Gerarde by about eight years, having been born in Marostica of Venice in 1553. He was at first a soldier, but pursued medical studies in Padua, where he earned his medical degree in 1578. Desiring to study plants, as well as the diseases and manners of other countries, he obtained in 1580 through his father's interest with the Senate an appointment to accompany the Venetian consul to Egypt. He remained in Egypt for three years, during which time he enthusiastically devoted himself to the study of plants and the medical practices of the Egyptians. He recounted his experiences in his book on the plants of Egypt,[27] in which he made the first attempt to explain the fructification of plants in scientific terms; in his dialogue concerning the celebrated Balm of Gilead;[28] and in his four books on the state of medicine among the Egyptians.[29] Such was the reputation acquired by Alpinus that upon returning to Italy in 1586 he was appointed physician to Andrew Doria, Prince of Melfi at Genoa. The Venetian Senate later appointed him as professor of botany and curator of the medicinal garden at Padua. He retained these appointments until he died in November 1616.[30]

Advancement of Knowledge

Upon looking back to the history of the sixteenth century and comparing it with preceding ages, we notice the vastness of difference between them. The faculties of man

[26] Catalogus Horti Johannis Gerardi, London, 1596.
[27] De plantis Egypti, libri tres, 4to. Venetiis, 1592.
[28] De Balsamo, Dialogus. Facc. 4to. Venetiis, 1592.
[29] De Medicina Egyptiorum, libri quatuor. Venet. 1611.
[30] Besides the works mentioned in the text, Alpinus wrote *De Rhapontico, Dissertatio Inauguralis*; *De Plantis exoticis*; *De præsagianda vita et morte ægrotantium*; published in 1601 and consisting chiefly of a collection and arrangement of the prognostics of Hippocrates; and *De Medicina methodica*, published in 1611, in which he endeavors to explain and revive the obsolete doctrines of the Methodic sect.

appear as if suddenly and violently awakened from an oppressive and anomalous torpor. We can attribute this mighty, almost incredible change to the discovery of the art of printing and its necessary result, the revival of letters, and to the emancipation of the human mind from the shackles of a gross and corrupting superstition. These were the glorious events that marked the end of the fifteenth century and whose effects began to appear in the sixteenth century.

Chapter 11

Medicine and Surgery in the Seventeenth Century

Codronchus

Early in this century, the town and vicinity of Imola were visited with a febrile affection that appears to have been produced by worms, of which an account has been given by Baptiste Codronchus, an intelligent physician of that place in a work[1] published in Bologna in 1603. He observes that only expulsion of a worm would allow relief, differing in appearance from those that are ordinarily found to infest the abdominal viscera. It is probable that, from the locality of the complaint and its having been a novel observation for the year, the worm that produced it had originated from ova deviant and more plentiful to the usual waters of that place. Codronchus was the author of several other ingenious medical works, which are found in the list of those recommended by Boerhaave in his *Methodus studii medici*. Among these was one on a safe and Christian mode of curing disease,[2] with a tract published from the close of the last century subjoined on the *Cocculus Indicus* and antimony. These berries, which are obtained from the *Cocculus lacunosus*,[3] a climbing shrub growing on the rocky shores of Celebes and the Moluccas, are employed by the inhabitants of the regions in which they grow to intoxicate fish and birds. In this country, brewers have had the credit of using them in the manufacture of their beer in order to communicate to it an intoxicating quality without the necessary expenditure of malt.

However, since an overdose might prove to be fatal, their use has been prudently prohibited by law. Yet in medicine they might perhaps be advantageously substituted in many forms for opium, in cases where the exhibition of that drug is either questionable or inadmissible. Besides these, Codronchus wrote papers on poisons and their antidotes, diseases

[1] De morbis qui Imoæ et alibi communiter anno 1602 vagati sunt Commentarius, in quo potissimum de lumbricis tractatur, et de morbo novo, prolapse nempe cartilaginis mucronatæ, 4to. Bononiæ, 1603.

[2] De christiana et tuta medendi ratione, cum tractatu de baccis orientalibus et antimonio, 4to. Ferrara, 1591.

[3] De Cand. Syst. Nat. vol. i. p. 519. It is from the root of another species of this genus, the *Cocculus palmatus*, that *calumbo* or *colombo*, one of the most valuable of our stomach bitters, is obtained. The *C. palmatus* is a climbing plant with a perennial root and annual stem and found frequently in the shady woods of the eastern coast of Africa between Obo and Mozambo, where it is regarded as a specific in the cure of dysentery and diarrhea, a fact noticed by F. Redi and J. Curvo Semmedo about the middlc of this century.

produced by witchcraft, hoarseness and other affections of the voice, hydrophobia, and hellebore as a medicinal plant.

Baronius

The second publication of merit that attracts attention in the seventeenth century is a quarto volume on affections of the urinary organs[4] written by Theodore Baronius, a physician of Cremona, who recommends the internal use of cantharides in such affections, a practice that Greenveldt probably learned from him. He also adopted the plan of curing calculus affections by means of lithontriptic liquors injected into the bladder. Baronius was a strenuous defender of the doctrines of Galen, maintaining that it was more honorable to err with him than to reason ever so correctly upon any other system. Fortunately, this blind and irrational bigotry did not mislead him in the path of practice, in which we find him frequently deserting his guide that would have led him into error.

Colle

The city of Padua, so distinguished for its university and the eminent talents of its several professors, especially in the various departments of medical science, was visited in 1610 by a pleurisy, which appears, from the account given of it by the professor of physic John Colle, to have partaken of the typhoid character, since bleeding at the arm did not admit of general depletion of the system, but readily improved under a course of local bleeding accompanied by a sufficient solubility of the bowels, which was produced by the use of mild purgatives. In a work on the practice of medicine,[5] which Colle published a few years after this visitation, he gives a history of epidemic diseases, wherein he includes an account of this pleurisy. In the year 1621, he published another work,[6] which was dedicated to Cosmo II and in which he lays down a very rational system of medicine and makes many excellent practical observations. In 1628, he published a paper on syphilis, in which he gives a concise history of the disease and its symptoms and lays down a plan of cure that involves using mercurial ointments and fumigation.

Colle was born in Belluno in 1558. He studied medicine in Padua under Cappivaccius and obtained a medical degree in 1584. He settled in Venice, and after practicing there for 15 years, he moved to Padua upon his appointment to the chair of medicine. He died in Padua in 1631.

---

[4] De operationis meiandi triplici læsione et curatione, libri duo, in quibus morbi renum et vesicæ, ex Galeni præsertim mente, pertractantur, 4to. Cremona, 1609.

[5] Medicina practica, sive methodus cognoscendorum et curandorum omnium affectuum pestilentium, fol. 1617.

[6] Cosmitor mediceus triplex, in quo exercitation totius artis medicæ decisa, ac consultations medicinales et quæstiones practicæ proponuntur, fol. Venetiis, 1621.

Van Helmont

About this period, the revolution that Paracelsus and his followers had commenced in the practice of medicine by the substitution of the chemical for the Galenical school of thought may be considered as having attained its completion under the daring lead of the celebrated John Baptist Van Helmont, who was the descendant of an illustrious family at Brussels, where he was born in 1577. In 1621, he published his first work[7] in support of the supposed property of the magnet in healing wounds.

Van Helmont appears to have exhibited early proofs of genius and such proficiency in his elementary studies that, by the time he was 17 years old, he had gone through the usual courses of philosophy at the University of Louvain and discovered how much hypothesis and how little real knowledge were contained in books of what were pompously called science. Such was his ardor in the pursuit of knowledge that before he was 20 years old he is said to have gone through the works of Galen twice, Hippocrates once, and nearly all the Greek and Arabian writers on medicine, and to have made copious commentaries on them. He appears to have formed a vague idea of the importance of experiment and induction as the only legitimate sources of solid information, but he does not seem to have methodized his ideas on the subject sufficiently to pursue that plan with full effect. Having attained his medical degree at Louvain, and being wearied with the unsatisfactory pursuit of knowledge in the pages of other authors, he commenced a course of travels for two years, during which he gave himself up, with the same zeal with which he had formerly devoted to the study of books, to the instructions of men who pretended to be acquainted with magic and philosophy and obtained acquaintance with the basics of chemistry from an illiterate man upon whom he stumbled. Having in 1609 married a lady of good family and large fortune, he retired to Vilforde, where he carried out his research uninterruptedly into the products of nature. There, too, the empirical knowledge he had acquired of the properties of several chemical remedies recalled his attention to professional practice, of which he had so long lost sight, and although he confined himself to giving advice to those who visited his residence for free, his advice was solicited by multitudes, and he boasts in his works of having cured thousands of individuals annually. In this manner he devoted his time to humanity and laboratory research for 30 years, of which resulted a considerable number of important discoveries in chemistry. In his work *De Humoribus Galeni*, he pointed out the gross absurdity of the Galenical hypothesis of the four elements. However, attempting like Paracelsus who preceded him and John Brown who followed him to generalize the undigested mass of facts that he had accumulated and reduce the whole of medicine to a few chemical principles, he substituted a jargon as unintelligible and theories as visionary as those that he had succeeded in expelling. However, he acquired at the time considerable reputation by the numerous works that he published and the discoveries in chemistry that he made. His doctrines continued for a long time to preponderate in the schools of medicine. His works are now, however, regarded only in the light of curiosities, containing a strange mixture of shrewd conjecture and absurd jargon. Some of his opinions may be found under new modifications in the speculations of later writers, and we can recognize his *Archæus* in a more refined form in the *Vis Medicatrix Naturæ* of Hoffman and Cullen, while Silvius and his followers adopted his theory of

[7] De Magnetica Vulnerum naturali et legitima curatione, contra Johannem Roberti Soc. Jesu Theologum, 8vo. Paris, 1621.

fermentations almost without alteration. His services were of the greatest value to chemistry, and he contributed largely by his efforts to prepare the way for modern discoveries. He died on December 30, 1644, at the age of 64 years.

Hornungius

In 1626, John Hornungius, Doctor of Philosophy and Medicine and chief physician to George Frederick, Marquis of Baden, published a collection of medical letters[8] written by some of the most distinguished men of the age on the subjects in which they each excelled. Among them are eight written by Gaspard Bauhin chiefly on subjects connected with botany, and three concerning matters in pharmacy from the celebrated Caspar Hoffman, professor of medicine at Altdorff. It is chiefly valuable for the light it throws upon the general state of medical science in Germany at the period of its publication.

Baronius and Bartholin

In 1636, Vincentine Baronius, a physician of Italy, published an account of that year's pneumonia epidemic. His description of the infection is full and accurate and displays great pathological judgment. He details with clearness and precision the most successful plan of treatment and gives a minute description of the postmortem appearances on dissection. This epidemic presented with cough, pain in the chest, dyspnea, fever, and inextinguishable thirst. When expectoration took place early, especially after bleeding had been carried to a due extent, the infection most commonly had a favorable termination. In fatal cases, the lungs were always found to be in an inflamed state, and serum was occasionally found poured out into the cavity of the chest.

The history of a case of aneurysm recorded by Thomas Bartholin in his *Anatomica Aneurismatis dissecti historia*, which was published in 1644, provides an instructive lesson to surgeons. It was a tumor in the leg that Antonius Vacca pronounced to be an aneurysm. The majority of opinions being against him, it was treated as an abscess, which allowed extension of the tumor to the toes. Gangrene ensued and led to the foot being amputated above the ankle. Three days later, they attempted to open the tumor, but the patient died during the operation. Although the artery was dilated six times its natural size, the part next to the skin was eroded and broken. A mass of blood similar to the consistency of flesh was found extravasated between the bellies of the gastrocnemius muscle. A similar case is recorded by Freind as having occurred at St. Bartholomew's.

[8] Cista Medica, in qua Epistolæ clarissimorum Germaniæ medicorum familiares et in re medica, & c. asservantur, 4to. Noribergæ, 1626.

Glisson and the Royal Society

Around the year 1645, Francis Glisson, who was Regius Professor of Physic at the University of Cambridge and a skillful anatomist, held weekly meetings in London for the purpose of discussing various scientific subjects. The attendees at these meetings at first were merely a few friends chiefly of the medical profession, but the number of attendees gradually increased and so did the regularity with which they attended these meetings. After the Restoration, the members were incorporated into an organized body under the title of the Royal Society. Glisson, one of the founders and original members of this society, had been appointed reader of anatomy to the College of Physicians, of which he was a fellow, and obtained much credit by his lectures *De Morbis partium*, which he published at the request of his colleagues. During the civil wars, he went to practice at Colchester, where he resided at the time of the siege in 1648. In 1650, he published his paper on rickets,[9] which exemplifies sound professional judgment and careful practical observation. His largest work[10] was one of a metaphysical character and appeared in 1672.

Badus, Bravo, and Chifflet

Peruvian bark had been recently introduced into the cure of fevers in Europe. It was met with opposition from the faculty and public opinion generally. Among those who took part in this controversy was Sebastian Badus, or Baldus, a native of Genoa who was one of its most strenuous advocates[11] and among the earliest writers in its favor. Badus had learned from a merchant of Genoa that the tree from which the bark was obtained grew in the neighborhood of Quito, and that it first acquired its celebrity and its name from having been employed in the cure of the Countess of Cinchon, for whom it was called Cinchona. Badus is extremely diffuse in his account of the cinchona, explaining at considerable length its properties and giving instances of its success in the cure of both intermittent and continued fevers. The works of Badus were controversial for the most part. He appears to have been a strong advocate for bleeding, which he recommended even in exanthemata.[12] He spent the latter part of his life at Rome under the patronage of Cardinal de Lugo, who was himself a great admirer of bark and had procured a considerable supply from Spain in 1649.

About the same time, Gaspard Bravo of Aguilar Campo in old Castile, physician to Philip IV of Spain and to the Inquisition, published a variety of works in which he mostly addressed the physiology and pathology of fever, the curative effects of bleeding, purging, and sudorifics, and the newly introduced Peruvian bark. Bravo was a graduate in medicine of the University of Valladolid, where he later became a teacher of medicine and surgery. In his consultations, he gives an account of the complaint that proved fatal to Philip with the manner of his death in 1665.

---

[9] Tractatus de Rachitide, sen morbo puerili *Rickets* dicto. London, 1650.

[10] Tractatus de Natura Substantiæ energeticæ, sive de vita Naturæ ejusque tribus primis facultatibus.

[11] Cortex Peruvianus redivivus, contra Plempium. 12mo. Genuæ, 1646, and Anastasis Corticis Peruvianæ seu Chinæ defensio contra ventilationem J. Jacobi Chifflet, et gemitus V.F. Plempii. 4to. Genuæ, 1663.

[12] Sanguis expiatus, seu de sanguine incalescente. Genuæ, 1663. Necessitas Phlebotomiæ in Exanthematibus. 4to. Genuæ, 1663.

John James Chifflet, a native of Besanç on, where he was born on January 12, 1588, and a physician of considerable reputation, was a strenuous opposer of the use of bark, against which he published a work in 1653 titled *Pulvis febrifugus orbis Americani ventilates*. In this work, he strongly condemns the exhibition of this medicine and says that intermittents stopped by it not infrequently return with increased severity.

Borelli

During 1647 and 1648, Sicily was nearly depopulated by the ravages of a malignant fever, accounts of which have been preserved by John Alphonso Borelli,[13] a native of Casteluovo, Naples, where he was born on January 25, 1608. Borelli, having been sent to Rome to complete his education under Castelli, made such rapid progress in his studies that the inhabitants of Messina invited him at an early age to visit Sicily as a teacher of mathematics, his studies having been directed to that branch of science no less than to medicine. He thus became a resident of Sicily during this destructive epidemic, and he enjoyed the most favorable opportunity to observe its progress, describe its symptoms, and decide on the most judicious method of treating it. Disgusted at length with his employers, Borelli left Messina in 1656 and accepted a professorship in Pisa, where his lectures met with the greatest success. He so ingratiated himself by his talents with the Grand Duke Ferdinand and Prince Leopold as to succeed, through their interest, in gaining his election to a seat in the Academia del Cimento. Here it was, in all probability, that he first conceived the design of accounting for the several functions of the animal economy upon mathematical principles, for which purpose he now diligently applied to the task of dissection, on which he wrote several letters. These letters may be found in Malpighi's posthumous works. In 1658, he published a second paper on the nature and treatment of malignant fever. At length, being supposed to have favored the revolters in an insurrection at Messina, he was obliged to leave the Neapolitan dominions, and, on an invitation from Christiana, Queen of Sweden, he moved to Rome, where she resided and where he passed the remainder of his days. However, he does not appear to have derived much pecuniary advantage from the royal favor, since we find him in the latter years of his life obliged to procure a scanty subsistence by teaching mathematics at the Convent of St. Pantaleon, where he died of an attack of pleurisy at the age of 72 years on December 31, 1679.

Patin

About this period, the celebrated Guy Patin, a physician distinguished for his learning and his wit, was elected dean of the faculty in Paris, and soon after succeeded the younger Riolan in the chair of medicine at the Royal College. He was the son of humble parents with contracted circumstances, residing at Hodene near Bray in Beauvais, where he was born in August 1601. Being employed as a corrector of the press in a printing office in Paris, he had the good fortune to form an acquaintance with Riolan, a physician of eminence and ability

[13] Delle ragioni delle Febri maligni di Sicilia. 12mo. Cosenza, 1649.

who, discovering the superiority of his talents, induced him to turn his attention to medicine. Patin obtained a medical degree in 1627 and was admitted as a member of the Parisian body. He continued to reside at Paris, where he obtained more celebrity by his wit than emolument by his practice, being of too proud and independent a spirit to fawn upon the great or pander to the appetites of the low. He acquired the friendship of many persons of eminent rank and distinguished talent. He was a zealous defender of the honors and privileges of his field and a strenuous opponent of all innovations in medical practice, specifically the introduction of Van Helmont's theories and drugs into the chemical school of physicians. In consequence of this aversion from novelties, he disturbed the tranquility of medicine in Paris by warmly entering into a dispute over the medicinal properties of antimony. He even attacked the personalities of those who advocated its use and drew up a long list of unsuccessful cases in which it had been employed, which he termed the *Antimonial Martyrology*. He believed that the judicious employment of the lancet, seasonable exhibition of purgatives, and administration of a few herbs were capable of effecting more substantial good in practice than all the studied and complicated prescriptions of the physicians. Patin appears to have planned a biographical history of the more distinguished physicians, but did not execute this project.

Patin died in 1672 at the age of 71 years, leaving a son, Charles, who was equally distinguished for his learning and talents both as a physician and as an antiquary. Although he was a voluminous writer on other subjects, he has contributed little to the stock of medical knowledge. A few out of his 40 different works in French, Italian, and Latin were on professional subjects, with the exception of the lives of the professors in Padua, consisting chiefly of detached papers. After travelling for some years, he settled with his family in Padua and was appointed to a professorship of medicine, from where he was transferred in 1681 to the first chair in surgery, and ultimately, in 1683, to that of medicine. He carried out his duties in so distinguished a manner that he was promoted by the Republic of Venice to the rank of a knight of St. Mark. He died in Padua on October 2, 1693.

The Hypothesis of Sylvius

Despite the great improvements made in other departments of medical knowledge, a circumstance occurred about the middle of this century, which tended materially to impede the progress of pathology. This was the introduction of the hypothesis of Sylvius, who, adopting with some slight modifications the chemical opinions of Van Helmont, attributed every morbid action within the animal economy to chemical changes that produced an acid-base imbalance. He imagined the whole of the cure to depend on the neutralization of this imbalance. Influenced by such fallacious reasoning and blind to the obvious and rational indications of cure exemplified by the disease's symptoms, Sylvius commenced his attack by the use of ammonia, absorbent earths, and cordials, or acids and refrigerants, accordingly as he imagined the acid or the alkali to predominate in the system. The extent to which this practice spread over Europe appears surprising, and the interruption to the improvement of medical practice and pathology was considerable. It is hard to say whether Sylvius did more to promote the cause of truth by his advocacy of Harvey's theory of circulation, or to impede it by his own rash and pernicious theory. In fact, it was one of the great benefits that Sydenham conferred upon medicine by his labors to expose the fallacy of such visionary assumptions and to guide practitioners from the regions of imagination onto the true path of legitimate inquiry, careful observation, and rational induction.

Chamberlen's Obstetric Forceps

An important improvement was introduced around 1660 into the practice of midwifery by the invention of an instrument for the purpose of facilitating delivery in cases of difficult parturition, the forceps. Contrived by Dr. Hugh Chamberlen, an eminent accoucheur in London, his father, and two of his brothers, they boasted that they were able by means of this instrument to deliver neonates with the most perfect safety and dispatch. Before this invention, it was usually necessary to sacrifice the life of the child in order to preserve that of the mother. Having established the reputation of his forceps in England, Chamberlen in 1672 proceeded to try his fortune with it in Paris, where, instead of adding to his fame, he encountered only disgrace by undertaking to deliver a woman whose pelvis was too narrow to admit the passage of the child's head without mutilation. The woman died in consequence of the attempt as predicted by Mariceau, an eminent Parisian accoucheur who was strongly opposed to Chamberlen's practice. To escape the disgrace of this failure, Chamberlen left Paris for Holland, where he is believed to have been more fortunate in his practice and to have imparted his secret for a valuable consideration to the celebrated Ruysch. He then returned to London, where he accumulated a considerable fortune, not so much through the success of his forceps, if we believe his adversary Mariceau, as through reading and translating Mariceau's *Observations sur la Grossesse*. Despite this sneer by Mariceau, which is derived more from jealousy than truth, Chamberlen's invention has been justly regarded as one of the most valuable instruments in midwifery, deservedly placing Chamberlen high among the improvers of his art. In 1683, Chamberlen published his translation of Mariceau's work, which was read with the greatest avidity.

Borri

Joseph Frances Borri, one of the most impudent and temporarily successful impostors any age or country ever beheld, was born in Milan around 1625 and educated in the Jesuits' College in Rome, where from his earliest youth he exhibited marks of a turbulent and licentious disposition; however, after posing as a devotee, he declaimed against the corruptions of the age and set up a claim to supernatural revelation. Afraid, however, of the power of the Inquisition, he withdrew from Rome and moved to his native town, where he became the leader of a new sect, binding his followers to secrecy by the strongest vows and depriving them of their property by imposing upon them the obligation of voluntary poverty. Arrogating to himself the credit of a divine mission, he took upon himself the episcopal privilege of conferring ordination by the imposition of hands. The arrest of some of his followers by the Inquisition awoke him from his visions of theocratic importance. Seeking security, he fled to escape the vengeance that overtook his deluded followers. He was publicly proclaimed to be a heretic. For failing to answer the charges brought against him, he was condemned for contumacy and sentenced to grace an *auto de fe* in effigy, which was done in

Rome in 1661. Meanwhile, he not only enjoyed security in Amsterdam against the familiars of the holy office, but also found in that town a new theater for the display of his matchless effrontery and daring impostures. Assuming the character of a physician and chemist, he claimed to possess extraordinary skill. To ensure his deceptions, he set up splendid facilities and assumed the title of "Excellency". However, his success in Amsterdam fell short of his expectations, so he thought it proper to suddenly leave at night, laden with all the gold and jewels that he could acquire, and shift to Hamburg, where he succeeded in imposing upon Queen Christina a belief that he possessed the secret of the philosopher's stone. After exhausting her means of supporting his extravagance and fearing an unwelcome detection of his impostures, he found it convenient to leave quickly and make Denmark the new theater of his exploits. Here he succeeded in levying contributions upon the credulity of the Monarch, until the death of the latter cut off this fertile source of supply and made it necessary for him to leave the Danish territories. While he hastened to astonish the Turks with a display of the versatility of his genius in Constantinople, he was mistakenly arrested on the frontiers of Germany as a political criminal. Here his good fortune abruptly deserted him when his name was accidentally reported to the Emperor in the presence of the Nuncio. He was claimed as a heretic and given up on condition of having his life spared, which was granted, but on reaching Rome he was condemned to abjure his errors, perform public penance, and be imprisoned for life. In prison, the fame of his adventures attracted a multitude of visitors of rank. After succeeding in restoring the health of the French Ambassador who had been given over by his physicians, he obtained a greater share of liberty and was permitted to continue his chemical pursuits in the castle of St. Angelo, where he died in 1695.

Introduction of Male Accoucheurs

About the same time that Chamberlen improved the art of midwifery by the invention of his forceps, an important revolution took place in the practice of the obstetric art by the introduction of employing male practitioners, whose assistance before then it had never been usual to call for in any but the most urgent and dangerous cases. This revolution was influenced in 1663 by the talents and reputation of Julian Clement, a well-known accoucheur who contributed largely to the improvement of the art. This distinguished practitioner was born in Aries in the Rhone in 1649. After receiving there a basic education, he went at an early age to Paris, where he was placed under the direction of James la Fevre, an eminent surgeon and accoucheur, whose daughter he married. He acquired la Fevre's practice when he died. By reason of his great reputation, in 1663 Clement was called to the Duchess de la Valliere, who, because of the novelty attending the introduction of male practitioners in such cases, is said to have moved to a private house for the purpose and received him in a mask, so that the rank or name of the person he attended to was concealed from him. After his success attracted the attention of the king, Clement was appointed accoucheur to the princesses of France. This appointment, in addition to his subsequent employment by other ladies of the court and his great popularity, succeeded in vanquishing injurious prejudices. In a little time, the complete substitution of male operators for female operators was effected in all obstetric cases. One of the most important results that followed this great revolution in practice was the progressive reformation of those abuses that had accumulated through ages of ignorance and timidity. Clement simplified the mode of turning the fetus in cases of wrong presentation and was the first to suggest the appropriateness of rupturing the membranes early in labor in cases of profuse and alarming hemorrhage, a practice successfully adopted by his pupil and assistant Parees. Louis honored him with a patent of nobility in 1684 on the express condition of his continuing to practice midwifery as long as his age and health would allow. He eventually died at the age of 80 years on October 7, 1729.

Arsenic in Cancer

The use of arsenic as a remedy for cancer to supersede the necessity of employing the knife for its extirpation was again brought forward in 1665 by Francis Blondel, a native and practitioner of Paris, in his epistle to Alliot. He employed in its cure a preparation of arsenicum rubrum, which was dissolved in nitric acid and precipitated by means of the superacetate of lead. The precipitate was then washed repeatedly in warm water, and its causticity further mitigated by burning spirit of wine, with which it was mixed until the precipitated powder became perfectly insipid. Blondel, who earned his medical degree at the University of Paris in 1632, acquired considerable reputation as a classical scholar and was employed upon the death of Chartier to assist in completing three unfinished volumes of his splendid edition of the works of Hippocrates and Galen. He was opposed to the use of antimony and all other chemical medicines, fully coinciding in opinion on this point with Guy Patin, his contemporary and coadjutor. We are indebted to him for an edition of the statutes[14] of the faculty of medicine at Paris, which he published in 1660.

Sydenham

In 1666, Thomas Sydenham (Figure 34) published the first specimen[15] of his revival of the inductive method when inquiring into the nature of disease, which Hippocrates had been the first to introduce with success, but of which, almost from the instant of his death, his followers had so unaccountably lost sight. Hippocrates' followers in many instances even corrupted his theories in order to obtain an apparent sanction for their own absurd deviations from the course he had so clearly and so judiciously pointed out as the only one capable of conducting them with true certainty.

[14] Statuta facultatis medicinæ, 12mo. Paris, 1660.
[15] Methodus curandi febres, propriis observationibus superstructa. London, 1666.

Figure 34. Thomas Sydenham (Image courtesy of Reynolds Historical Library, the University of Alabama at Birmingham).

Sydenham, on applying to the study of medicine, early detected the fallacy of the theories usually inculcated in books, and held the medical works of his time and lessons taught at the principal medical schools throughout Europe in great contempt. While it led him at a glance to detect the fallacies of the existing systems, his penetrating genius also conducted him into a more rational mode of acquiring correct notions on the subject of pathology. In the dedication of his *Medical Observations on the History and Cure of Acute Diseases*, he informs us that upon commencing practice he became immediately convinced that the only means of acquiring a correct knowledge of the fundamental principles of his profession consisted of attentively observing the whole disease phenomena and inquiring into the progress and fluctuations of the several symptoms. The true and natural indications of cure thus could be deduced. He first applied this inductive method to febrile diseases, and only after many years of close attention did he fully satisfy himself as to their proper and judicious mode of treatment. As he elegantly expresses himself in the preface to his *Medical Observations*, in writing a history of diseases the writer should dismiss from his mind every theory that might prejudice his judgment and make note of, in the most accurate manner possible, all of the natural appearances of disease. He commences with the Hippocratic definition of disease as a severe effort of nature to expel the matter in order to restore constitution. Thus, he considered plague to be an effort of nature to expel the humors by means of buboes. However, he appears in his practice to have discarded the deceptive light of hypothesis and professed nature as his only guide. He moderated the eruptive fever of smallpox by cool air remedies, by which the subsequent eruption was greatly lessened and the danger proportionally diminished. The adoption of this plan contradicted his hypothesis, according to which, as a salutary effort on the part of nature to relieve the constitution, the increase of the eruption ought to have been promoted. However, here, as in other instances, Sydenham furnished a useful example of the triumph of common sense over speculative delusion. Modern experience has fully corroborated the wisdom of his plan. The histories that Sydenham has left us of measles, gout, smallpox, and other disorders afford striking evidence of the depth of his wisdom and the soundness of his observations, and are justly regarded as models of what medical histories should be. The influence that difference of season exerts over disease in modifying its character or treatment did not escape his penetration. He has even extended his observations to what he terms the epidemic constitution of different years, which he regards as likewise modifying the character of the prevailing complaints.

Sydenham published many works.[16] His paper dedicated to Dr. Short on gout and dropsy and his observations on calculus are particularly deserving of careful study. Sydenham suffered from both gout and calculus. He is decidedly opposed to venesection, whether with a prophylactic or a curative intention, to treat gout. He objects equally to the system of purging recommended by so many practitioners both of his day and of later periods, and to the system of periodical purgation adopted by many. He regards this latter system as frequently augmenting, instead of diminishing, the frequency and severity of the paroxysms. He also objected to the plan of diaphoretics, regarding evacuations of all sorts that were carried beyond the degree necessary for health as injurious. He rested his curative plan chiefly upon a class of medicines that appear little suited to a complaint of an inflammatory character, medicines that he says gently warm the system and are either slightly pungent or leave a sensation of bitterness upon the tongue. Sydenham himself sank under a continued succession of attacks from gout and calculus and died in December 1682 at the age of 65 years.

Lower and Denys

In the *Philosophical Transactions* for 1666, we meet with the first account of experiments made upon the transfusion of the blood of one animal into the veins of another, which was laid before the Royal Society by Dr. Lower, whose speculations probably suggested a repetition of the experiments that he had made upon animals upon the human body. This was attempted, probably for the very first time, by John Denys, a graduate of the University of Montpellier. Denys published the result of his observations in the *Journal des Savans* for 1667. Denys was an enthusiast on the subject and confident of success from the plan that he recommended. In a letter[17] to M. de Montmor, he tells him that by these means he restored a lunatic to his senses and cured a Swiss gentleman of the name of de Bonde. However, his patients falling victims to his practice, despite his assertions of their cure, further experiments upon human subjects were expressly prohibited by the parliament. Some attempts have been made in the nineteenth century to revive Dr. Lower's and Dr. Denys's practice, but this attempt to invade the patent rights of that illustrious pair is not likely to prove very successful or fashionable anywhere beyond the precincts of St. Luke's.

[16] Hamilton lists several of Sydenham's works in the original text.

[17] Lettre écrite à M. de Montmor touchant une nouvelle manière de guérir plusieurs maladies par la transfusion du sang. 4to. Paris, 1668.

Cole and Archer

Among the other believers in the efficacy of Cinchona we find William Cole, a learned and ingenious physician. After having been educated at Oxford, where he took the degree of bachelor of medicine in 1660 and that of doctor on July 7, 1666, he settled in Bristol. He then moved to London. In a work[18] that he published at Oxford on the animal secretions, he ascribes the peculiarities of their quality to a peculiarity of structure in the secreting glands. In 1689, he published an *Essay on the Frequency of Apoplexies*. In 1694, he published his new theory on fever, which he ascribed to a vitiated state of the nervous fluid. In another work,[19] he strongly recommends the use of the mistletoe in epilepsy and pretends to have cured the disease, inserting a history of the case given by the patient himself, in which he says he was very much relieved.

A much more useful work than Dr. Cole's treatise on the medical or the magical properties of the mistletoe was a manual published in 1673 by John Archer, *Every man his own Physician*, which, although little noticed, was a useful work for its day. From a list of his inventions subjoined to his work, Dr. Archer appears to have been a man of considerable ingenuity, having anticipated some of the more remarkable of our modern improvements. One of Dr. Archer's inventions was a vapor bath for rheumatism.

Buschoff

In 1675, Herman Buschoff, a native of Utrecht but resident of Batavia for many years, introduced from India a new method of curing gout by the application of fire. During his stay there, he had been induced while laboring under an attack of gout to hire an Indian woman to cure him, which she did by burning little pellets of moxa on the part affected, and found such relief from this mode of applying the actual cautery that he was induced to collect and author a short paper written in Dutch on a number of cases illustrative of its efficacy. He sent this paper to Holland to be printed, accompanied by a specimen of the moxa and the sandalwood matches that were used to light the pellets. In his work, he speaks of the frequency of this mode of treatment among the Chinese and Japanese not only in gout but also in other conditions. He says the operation lasts about half an hour and causes only minimal pain, leaving small scars that resolve in a few days and little eschars that heal with any common application. After his own operation, he says he fell into a sound sleep for 24 hours and awoke perfectly free from pain and swelling. Upon experiencing a second attack of gout two years later, he had recourse to the same remedy with the same success. He also stated that he had known this remedy to be employed with equal success in the cure of schirrous tumors, nodes, and lethargies. At the time that Buschoff's treatise appeared in Holland, Sir William Temple, who was resident as British Ambassador at the Hague and also suffered from gout, was easily convinced to try the new remedy. He was so pleased with the effect that he wrote his own paper on the subject, explaining the manner in which the moxa was to be used and the benefit that he had experienced from it. Buschoff's work was also translated into English in the same year at the instigation, as we are told by Sir William, of the professors of

[18] De secretione animali cogitate, 8vo. Oxon. 1667.

[19] Consilium ætiologicum de casu quodam epileptico.

Gresham College, of whom no notice, however, was taken in either the title or the preface. For a time, the moxa was a fashionable remedy, but eventually its effects fell short of the extravagant expectations that had been formed of it, and it fell into disuse.

Blasius

Gerard Blasius, a Dutch physician of some repute, published in 1687 a small volume of medical observations,[20] which contains an account of various morbid appearances and deformities of the body. Many of these cases are extremely curious and serve to mark the growing improvement of medical observation and inquiry. He mentions the case of a 12-year-old boy who had been operated upon for the stone in March 1675 and who over a period of six months passed upwards of 100 worms. In the succeeding observation, he gives an account of seeing two worms of great length and unusual structure in the kidney of a man. Of these unique worms, figures are given in the eighth of the inserted tables, which exhibit a manifest improvement in the art of delineating objects of natural history. Among the cases of anomalous structure, he notices a deficiency of the right kidney in one instance and of the urinary bladder in a second. In a third, he speaks of two stomachs, and in a fourth of three kidneys. Blasius was the author of several other works on anatomy, both human and comparative. In a letter inserted in the third century of Bartholin's epistles, he claims to have discovered the salivary duct, which has been erroneously assigned to Steno, but truly belongs to him, he having first shown it to Steno, whose name it improperly bears.

Young

In 1678, an improvement was made in the practice of amputation by Mr. James Young, a surgeon of Plymouth. He placed a hard bolster of linen on the inside of the limb in the direction of the main artery, passed a bandage around the limb, tied the ends together, and twisted them so tight as to interrupt the circulation effectually. He recommended loosening the tourniquets a little before the dressings were applied in order to discover where astringents were most needed. He gave an account of this plan in a scarce book published in London in 1679, *Currus triumphalis e Terebintho*, in which he proposes amputating with a flap, since claimed as a discovery by two French surgeons, Verduin and Sabourin.

In 1679, Stephen, son of Nicholas Blancard, published his *Medical Lexicon*, a useful and laborious compilation. This work contains the explanation of many medical terms, such as the student may experience some difficulty in finding elsewhere.

[20] Observationes medicæ rariores. Amstelodami, 1667.

Blegny and the Petrified Child

An account[21] published in 1679 by Nicholas Blegny, a quack rupture doctor, discusses a case in which a fetus was said to have been petrified, having acquired from its long residence in the abdomen and the pressure of the other viscera an almost cartilaginous hardness. Had Blegny been acquainted with the doctrines of physiology, he would have known that simple pressure was more likely to produce absorption than petrifaction. The whole account comes before us in so questionable a form that although the case already related, upon the authority of Bartholin and Paré, justifies our admitting the possibility of some real foundation existing for Blegny's report, implicit reliance should not be placed on all the statements of his narrative.

Blegny had previously published a small volume on the art of curing ruptures,[22] which has been frequently reprinted. His method was to cauterize the skin of the groin with nitric or muriatic acid. After the ensuing wound had healed, a firm scar remained, and the scar helped resist additional descent of the intestine. It was the invention of the Prior De Cabeveres, and he instances several remarkable cures that he effected by this method. In one of the cases that he relates, the urinary bladder had slipped into the ring.

Blegny also authored a work on syphilis,[23] which was known, as he pretends, to Moses and the ancients, and may be produced by immoderate venery. He justly censures the use of astringent injections in cases of gonorrhea, and professes to cure lues with more certainty and safety by using guaicum and sarsaparilla instead of mercury.

Bellini and Boerhaave

In 1683, Laurence Bellini, who was born in Florence in 1643, published a quarto volume on the subject of the pulse and urine, venesection, fevers, and disorders of the head and chest,[24] of which a fourth edition was published at Leyden in 1717 with an attached preface by the celebrated Boerhaave. It is dedicated to Francis Redi and is a work of considerable observation and research, although often obscure and considerably too theoretical. Bellini's work contains an excellent description of that rare affection known as catalepsy, in which all voluntary powers are suddenly arrested and the body becomes fixed throughout the spell in the same attitude in which it commenced. To use the emphatic language of Bellini himself, the body was suddenly converted into "rigid iron, or inflexible wood",[25] or perhaps, in stricter language, the patient suddenly becomes like a breathing statue, incapable of thought, volition, or spontaneous motion, while at the same time the vital or involuntary functions, those functions that the soul or brain has no control in eliciting, such as the respiration and circulation, go on unimpaired, apparently unaffected by the apparent absence of the soul.

---

[21] Histoire anatomique d'un enfant qui a demeuré vingt cinq ans dans le ventre de sa mere, 12mo. Paris, 1679.
[22] L'art de guérir les hernies. 12mo. Paris 1676.
[23] L'art de guérir les maladies vénériennes, expliqué par les principes de la nature, et de la mécanique, 2 vols. 12mo. Paris.
[24] De urinis et pulsibus, de Missione Sanguinis, de Febribus, de Morbis Capitis et Pectoris. Opus Laurentii Bellini, dicatum Francisco Redi. 4to. Bononiæ, 1683.
[25] De morbis capitis. p. 593. Lugduni Batav. 1717.

Closely connected with the speculations to which the description of catalepsy almost unavoidably gives birth is the subject assumed by the celebrated Boerhaave for his inaugural thesis upon earning his degree in philosophy in 1690. When taking for his theme *De distinctione mentis à corpore*, he ably refuted the errors of Hobbs and Spinosa. However, being suspected of a latent attachment to the very doctrines whose fallacy he had so completely exposed in his thesis, he was induced to abandon his original intention of treading in the footsteps of his father as the pastor of a village church and, directing his attention from theology to medicine, became one of the brightest ornaments of the profession. Chemistry was the branch of study that appears to have principally engaged the young Boerhaave's attention. However, he did not devote himself exclusively to any one branch. He studied the works of Vesalius, Fallopius, and Bartholin, and learned dissection under Nuck. Concerning figures in medicine, he preferred Hippocrates among the ancients and Sydenham among the moderns. In 1693, he obtained his medical degree at Harderwyckin in Guelderland at the age of 25 years. The remaining particulars of his brilliant career belong to the history of the succeeding century.

Amman

About this time, the situation of the deaf and dumb began to attract the attention of practitioners. John Conrad Amman, a native of Schaffhausen and graduate of the University of Basil, from where he obtained his medical degree in 1687, applied himself particularly to the discovery of a method by which persons unfortunately deprived of both hearing and speech from birth might be taught to express themselves intelligibly, or at least make their wants known. In this useful and humane endeavor Amman appears to have been eminently successful, as is fully evinced in his treatise on the subject, which was first published in Dutch in 1662 under the title *Surdus loquens* and later translated into Latin and republished in 1702 with the additional title of *Sett dissertatio de loquela*. Amman's work has steadily maintained its original reputation, gone through numberless editions, and been translated into almost every European tongue.

Colebatch

In 1695, John Colebatch, a surgeon and apothecary of much talent and industry, published an octavo volume, *A new light of chirurgery*, in London. In this work, he points out the mischief of using tents and injecting acrid substances into wounds. He advises the substitution of a powder of his own composition dissolved in water, which he claims halts hemorrhage, allays pain, and disposes the parts to heal. Naturally, his plan excited considerable opposition, so he republished it a few years later, accompanied by an extensive vindication and a variety of cases illustrative of its success. In 1696, he published *A Physico-Medical Essay, concerning alkali and acid*, in which he adopted the chemical principles of Van Helmont and Sylvius, attributing almost every complaint to a redundance of alkali. He maintained that the most rational plan of cure consisted in the application of acids, such as lemon juice, cream of tartar, and sulfuric acid. These were his grand panaceas, especially in the cure of gout, on which he published a treatise the following year. His last publication was on the medicinal properties of the mistletoe, the efficacy of which in epilepsy, chorea, and other convulsive disorders he gave numerous instances from his own experience.

Belloste

Contemporary with Colbatch was Augustin Belloste, a native of Paris and the inventor of a mercurial composition named after him, "Belloste's pill", by which he is believed to have accumulated a considerable fortune. Its composition was made a mystery, which renders its merits impossible to speak of. Belloste, as well as Chamberlen, the inventor of the obstetric forceps, set an unworthy example by concealing the preparation of a composition calculated to lessen human suffering out of the mercenary view of turning the distresses of his fellow creatures to his own private advantage. That Belloste was capable of better things and influenced at times by nobler feelings has been sufficiently proven by the publication in 1695 of his *Chirurgien de l'Hopital*. A few years later, he published a continuation of this work, in which he recommended with Celsus the piercing of carious bones in order to promote their exfoliation, a practice that has long fallen into disuse. Belloste says that the practice of changing dressings of wounds too frequently retards their healing.

Baglivi

George Baglivi was born in Ragusa in 1668 and appointed professor of the theory of medicine and anatomy in Rome by Pope Innocent XII. He published his first work, *De Praxi Medica ad priscam observandi rationem revocanda*, in 1696, in which he deplores the low condition of medicine of his time. He charges to the neglect of observation and experiment, the omission of studying the ancient Greek writers, Hippocrates especially, and an inordinate fondness for speculative rather than experimental reasoning. He admits, nevertheless, the progress made in anatomy and physiology, and the superiority of the modern theories to the dreams of the ancients, and anticipates that when the moderns shall sedulously apply themselves to practical observation, they will surpass the ancients as far in their knowledge of the true treatment of disease as they then did in the soundness of their theories. Inquiring whether theory or practice is more conducive to knowledge of the proper treatment of disease, he decides in favor of practice, yet at the same time he recommends a combination of both.

He was also accused of plagiarism and censured for credulity in allowing himself to be influenced by knaves pretending to suffer from various affections from the bite of a tarantula that only admitted of cure by music. However, on this point we should be lenient to Baglivi, since we find even the enlightened Englishman Mead half a century later attempting to account for this supposed phenomenon upon philosophical principles. In 1696, Baglivi published his dissertation, *De anatome, morsu, et effectibus Tarantulæ*, to which succeeded his work *De Fibra motrice et morbo*, which contains his theory concerning the origin of the motion of the solids, which he ascribes to a consent between the heart and dura mater. He has been accused of borrowing this theory from Paccioni, but he says that he himself communicated it first to Paccioni. Baglivi was the author of several other works, all of which were collected and published under the title of *Opera Omnia*, which will always merit the attention of the medical student.

Cyprian

Abraham Cyprian was a native of Amsterdam and went to Utrecht for his education. He established a practice in Amsterdam and resided there until 1692, leaving for Franknaer to serve as the chair of anatomy there. In the last year of this century, Cyprian published an account of an extra-uterine fetus taken from one of the fallopian tubes, in which it had lain for 21 months. Although the woman recovered from the operation, her death a few years later furnished Cyprian with an opportunity to examine her body and thereby discover the place in which the fetus had been lodged. Cyprian was also an eminent lithotomist, having operated on at least 1,400 patients. He has left an account of his method of operating in his *Cystotomia Hypogastrica*, which was published in 1724.

Chapter 12

Anatomy and Physiology in the Seventeenth Century

Discovery of the Lacteals in 1622

The discovery of the lacteals, an important prelude to that of the circulation, was made by Gaspard Aselli in 1622 by dissecting a dog that had just eaten a meal, by which means the lacteals of the mesentery were rendered distinctly visible as they ran across it, distended with the milky fluid of the newly absorbed chyle. However, Aselli, ignorant of their true course, mistakenly described them as passing from the intestines to the liver and confounded them with the lymphatics of that internal organ. He admitted that the lacteals had been acknowledged by some of the ancient writers, although they had neither been described nor had their functions demonstrated. Among the anatomists of this century, only Aselli anticipated the importance of the lacteals. Aselli himself made errors in his endeavors in anatomy. For example, he mistook a collection of glands in the mesentery for the pancreas and described them as a new discovery.

Six years after the discovery of the lacteals, William Harvey (1578-1657), who had been led by Fabricius ab Aquapendente (1539-1619)[1] to investigate the long disputed problem of the circulation of the blood, solved the mystery that had so long baffled the minds of the greatest men. Harvey published his theory of the circulation in an essay published in 1628.[2] Harvey (Figure 35) was born in Kent, England. He was initially trained at Canterbury and Cambridge, and then went to Padua. After graduating from medical school in 1602, he returned to England and became a physician to King Charles I. Harvey died from a stroke at the age of 79 years.[3]

Harvey, following up on the discoveries of which his instructor was unable to understand the application,[4] elucidated the mechanism of the circulation (Figure 36). He demonstrated that the blood does not merely undergo its minor circulation through the lungs, but flows in

---

[1] See, McAlister VC. William Harvey, Fabricius ab Acquapendente and the divide between medicine and surgery. Can J Surg 2007; 50, 7-8.

[2] Exercitatio anatomica de Motu Cordis et Sanguinis in Animalibus. Francf. 1628.

[3] See, Androutsos G, et al. William Harvey (1578-1657), discoverer of blood circulation. Hellenic J Cardiol 2012; 53, 6-9.

[4] This probably refers to Fabricius ab Aquapendente's observations of the valves of the veins.

one full stream from the left ventricle of the heart through the aorta, and then through numerous arterial branches to every part of the body, and after performing its functions, it is taken up by the various branches of the veins, pours into the vena cavae,[5] and then returns to the right atrium and ventricle of the heart to prepare for the inhaling of air in another passage through the lungs before resuming its circle through the system.

Figure 35. William Harvey, 1578-1657. "Guilielmus Harveius, Colleg. Medicor Londin. Socius" (Image courtesy of Reynolds Historical Library, the University of Alabama at Birmingham).

[5] Notably, Cæsalpinus believed that blood oscillated in a perpetual ebb and flow through the same vessels.

Figure 36. William Harvey, 1578-1657. "Harvey Demonstrating the Circulation of the Blood." (Image courtesy of Reynolds Historical Library, the University of Alabama at Birmingham).

Discovery of the Pancreatic Duct in 1642 and the Lymphatics in 1651

The pancreatic duct was discovered in 1642 by Johann Georg Wirsung (1589-1643), a Bavarian anatomist who was assassinated by an Italian student supposedly hired to kill him. The reason for Wirsung's assassination is not clear, but it is thought to be related to "jealousy over the discovery of the main pancreatic duct."[6] Wirsung studied under Johann Vesling[7] (1598-1649) in Padua.[8] Almost at the same time, three unacquainted anatomists living in three separate countries, Sweden, Denmark and England, appear to have discovered the lymphatic vessels. The credit of this original discovery seems to belong equally to all three anatomists, but in point of priority, we believe the credit goes to the Swedish anatomist. As early as 1649 or 1650, Olof or Olaus Rudbeck (1630-1702), son of the Bishop of Vesteras in Sweden, discovered the lymphatics of the liver when he was only 19 or 20-years-old, and soon after, discovered lymphatics in other parts of the body. He is thought to have given some account of this discovery in his inaugural thesis,[9] the subject of which is Harvey's doctrine of the circulation of the blood. The discovery of Bartholin of Denmark, which appears to have been perfectly independent of that of Rudbeck, was first published in 1653, nearly at the same

[6] Giardiello C, Trojaniello B, Giardiello A. Murder for a duct. Wirsüng and Hoffmann, the true story of the discovery of the main pancreatic duct [Italian]. Chir Ital 2007; 59, 857-60.

[7] Vesling was the author of *Syntagma Anatomicum*, which was translated into many languages and became a landmark anatomy book for its illustrations. See, Maxwell H, Selwyn S. A History of Medicine. New York, Knopf, 1947.

[8] See, Beger H, et al. The Pancreas. Oxford, Blackwell, 2008.

[9] Vasa lymphatica nuper Haffniæ in animalibus inventa et in homine, 4to. Haffniæ, 1653.

time that the same discovery was made by the English physician Jollife without the slightest acquaintance with the discoveries either of Rudbeck or Bartholin. Haller credits Rudbeck for discovering the lymphatics, but he fully admits the claim of Bartholin of having traced and described the course of these vessels with the greatest accuracy.

Discovery of the Thoracic Duct

The discovery of the lacteals on the part of Aselli was complemented by Nicholas Pequet's tracing these vessels to their common receptacle, the thoracic duct. Pequet was a native of Dieppe. Anatomists were misled by the erroneous description of the lacteals by Aselli and imagined that the lacteals terminated their course in the liver. The discovery of the thoracic duct had been made nearly a century before in 1563 by Eustachius, but he by no means understood the nature or importance of its functions. Thus, his description was little regarded before Aselli's discovery of the lacteals. Before and for upwards of a quarter of a century after Pecquet's discovery of the thoracic duct, Aselli and other anatomists, believing in the erroneous opinions of Hippocrates and Galen, imagined that the mesenteric veins absorbed the chyle from the intestines and conveyed it to the liver, instead of pouring it, as Pecquet described, into the circulation for its appropriate nourishment of the body. Therefore, the discovery of the thoracic duct must be indisputably credited to Eustachius, whereas the first description of valves within the veins is credited to Fabricius. The credit for connecting the discovery of Aselli with that of Harvey by demonstrating the functions of the thoracic duct belongs to Pecquet. Despite these discoveries, the question as to the identity of the lacteals and the lymphatics remained undecided and until the following century, when experiments carried out in the school of the Hunters finally established a satisfactory distinction between the two systems.

Discovery of the Cystic Duct

Francis Glisson was a professor of physic at Cambridge and discovered the cystic duct. His work on the anatomy of the liver, which was published in 1654, gave a more accurate description of that organ than had previously appeared. He is also credited with the discovery of the capsule of the liver, which bears his name. However, it was previously noted by Pecquet and Walseus, but Glisson was the first to examine and describe it accurately.

Discoveries in Oral Anatomy

The English physician Thomas Wharton published an essay on the glands titled *Adenographia* in 1656. This essay contains a number of new details about these organs, which added considerably to the stock of anatomical knowledge. In a work on the anatomy of the human frame[10] published in 1657 by Nathaniel Highmore, we find several discoveries that

[10] Corporis humani disquisitio anatomica, Fol. Hagæ Comitum, 1657.

do not belong to him but for which he has been credited. For example, the maxillary sinus, which has been named the *Antrum Highmorianum* in his honor, was noticed before him by Casserius, who referred to it as the *Antrum geno*. Highmore was the first to suggest extracting a tooth in the setting of a cavitary abscess, and this procedure was later practiced by Cowper. Highmore was a graduate of Oxford, where he earned his medical degree in 1642.

In 1662, Nicholas Steno of Denmark described the salivary duct belonging to the parotid glands as his own discovery. The duct has been named after him, although Blasius, in a letter to Bartholin, claims the discovery for himself and says that he first pointed out the duct to Steno.

Malpighi

Malpighi, or Marcellius Malpighius (Figure 37), was a prominent researcher. Through his microscopic investigations, he detected the glandular structure of the cortical portion of the brain and pointed out the true nature of the tongue. To him we are indebted for a correct knowledge of the cellular disposition of the lungs and the formation of the substance of the liver.

Figure 37. Malpighi (Image courtesy of the National Library of Medicine).

He also examined and described the structure of the spleen. We also credit him for providing the first accurate description of the kidneys, from which we learn of their lobular structure, even though we may disagree with his opinions concerning their final and minute organization. Malpighi taught that the glandular acini were hollow for the reception of the fluid secreted by the arteries distributed on their surface, from where it is conveyed by an excretory duct, and that the union of a number of these ducts forms the larger excretory canal. Ruysch denied this principle of Malpighi and maintained that these excretory ducts were merely continuations of the same tube. Malpighi regarded the mucous glands of the alimentary canal as examples of the simplest form of a glandular structure.

Ruysch

In the year 1665, the distinguished anatomist Frederick Ruysch, who was known for introducing colored injections within anatomical preparations that gave them the freshness and appearance of life, published an essay[11] on the lacteals and lymphatics, in which he discussed the results of his inquiries during a controversy in which he had been engaged with Sylvius and Van Home. His talents had been sought by these anatomists to combat the vanity of Bilsius, who came to Leyden while Ruysch was still only a student to exhibit his boasted method of making anatomical preparations. In this work, in defense of Sylvius and Van Home, Ruysch does not deny the previous description of the valves of the lymphatics, but claims to have first demonstrated the method in which one discovers them. The fame of this work procured him an invitation to fill the chair of anatomy at Amsterdam in 1666, and he accepted the position. The discovery of the injections that Ruysch so successfully employed in making those universally admired preparations originated with De Graaf and was improved by Swammerdam. Ruysch deserves the credit of having turned this method to more practical use. He prepared the entire bodies of infants and adults with this technique. It is not surprising that Ruysch's museum should have been the most magnificent ever possessed by a private individual or that he made discoveries that had eluded other anatomists. He thoroughly examined the pulmonary circulation and claimed to have discovered the bronchial artery. He also examined the structure of the brain, ear, and lymphatic and glandular systems. In 1685, Ruysch was appointed professor of physic and held this position until 1728 when he fell and fractured his thigh. He held the office of midwife superintendent and abolished the practice of speedily extracting the placenta, which he believed to be expelled by means of an orbicular muscle at the fundus uteri. He died in 1731 from a fever at the age of 93 years.

Meibomius

Henry Meibomius was a skillful anatomist who was born at Lubeck in June 1630. He received his medical degree from the University of Angers. In 1666, he published a letter to Langelot[12] in which he describes some vessels of the eyelids that he had recently discovered. From a letter published and addressed to Velschius, he appears to have planned on writing a history of medicine, but the difficulties he encountered in collecting the necessary materials seem to have deterred him from proceeding with that project. In 1664, his edition at

[11] Dilucidatio valvularum in venis lymphaticis et lacteis, cum figuris æneis, Hagæ Comiti, 1665.
[12] De vasis palpebrarum novis. Epistola ad virum celeb. Joelem Langelotum. Helmstadt, 1666.

Helmstadt of Arnold de Boot's *Observationes medicæ de affectibus omissis*, which had been previously published in London, derives an additional value from the notes and preface with which he enriched it.

Needham, Borelli, and Swammerdam

Walter Needham, an English physician, gave an account of the membranes that envelope the fetus in a work[13] published in 1667. In 1669, Borelli published a work aimed to show that the powers of the eyes are unequal in most men, with objects appearing much more distinct with one eye than the other. In 1670, his commentary *De motionibus naturalibus a gravitate pendentibus* appeared posthumously as a prelude to his great work on physiology *De motu animalium*, the first volume of which appeared in 1680 and was dedicated to Queen Christina. The object of this work was to explain the animal functions upon mechanical principles. John Swammerdam was an enthusiastic anatomist and proficient entomologist in early life. While residing in Leyden, he distinguished himself in anatomy and exhibited singular skill in making preparations. He was a fellow pupil of Regnier de Graaf under the celebrated anatomy professor Van Horne. Before earning his degree at Leyden, Swammerdam went to Paris for further training and there met Steno. His thesis for his degree was on respiration. At this time, he began to employ his composition for injecting subjects and thereby render visible the arteries and veins of the face. Religious scruples prevented him from continuing to use this valuable invention, but his vanity, however, led him to communicate his discovery to Ruysch, who implemented this technique in the course of his career. In 1672, he published his work[14] on the female uterus. Intense application made him a hypochondriac unfit for society, and in this state he became so impressed with the reveries of Antoinette Bourignon that he plunged into the depths of mysticism. He died in 1680 at the age of 43 years.

Regnier de Graaf

Regnier de Graaf was the son of an architect residing at Schoonhove in Holland. Regnier was born on July 30, 1641. He began his studies at Leyden, but graduated from medical school in Angers around 1665. At the age of 22 years, he had published a dissertation on the subject of the pancreatic juice. Upon returning to Holland he settled in Delft, where his practice was successful. He published three dissertations on the organs of reproduction in both sexes, which involved him in a controversy with his former classmate Swammerdam. He died in August 1673, at the early age of 32 years, leaving behind several esteemed works that were published in Leyden in 1677 and 1705.

De Graaf developed those injections that Swammerdam brought to perfection and that Ruysch employed to so much advantage. De Graaf was first led to employ these injections for the purpose of discovering the motion of the blood in the vessels. For this purpose he contrived a new type of syringe with which he was able to inject the vessels with a high-colored substance and thereby make the course of the circulation clearly demarcated.

[13] De format fœtus, 8vo. Lond. 1667.

[14] Miraculum naturæ, seu uteri muliebris fabrica notis in J.V. Horne prodomum illustratum, 4to. Lugd. Batav. 1672.

However, De Graaf's injection was too thin to keep the vessels permanently distended, and thus this technique fell into disrepute until Swammerdam observed the cause of the defect and proposed to remedy it.

Briggs

In 1674, William Briggs, a native of Norwich and fellow of Bennet College at Cambridge, published a paper on the eye,[15] in which he pointed out his discovery that in the tunica retiformis, which is contiguous to the vitreous humor, the expanded filaments of the optic nerve lie in the most regular order and continue to later retain this order when afterwards united in the nerve, and preserve it unbroken until they reach the brain. The chrystalline humor had been previously discovered to be a double convex lens, formed of two segments of unequal, but not perfectly spherical, spheres. Briggs' discovery explains why all the images are carried so distinctly to the brain. Briggs also described in this work the ducts by which moisture is conveyed from the glands in the angles of the eyes to lubricate and facilitate motion within the orbits.

Genga and Peyer

Bernardine Genga, the author of an anatomical work[16] published in Rome in 1675, was a strenuous advocate for Harvey's theory of the circulation, the discovery of which, however, he stoutly maintained to have been made by Father Paul Sarpi, although he produces no adequate evidence with which to support this claim. However, through his exertions, Harvey's doctrines gained ground in Italy. Genga charged Hippocrates with committing errors in the treatment of several surgical cases. The first accurate account of the glands was given by John Conrad Peyer, a native of Schaffhausen, in a work[17] first published in 1677. He also gave an account of some experiments he made upon the pancreas in a tract contained in the *Bibliotheca Anatomica* of Le Clerc and Magnetus.

Bonetus, Schellhammer, and Blancard

A valuable work on morbid anatomy discussing a vast number of dissections of persons who had died of disease uncovered many hidden causes of illness and was published in 1679 in Geneva by Theophilus Bonetus. Some useful observations on the tongue, salivary glands, larynx, diaphragm, intestinal canal, mesentery, thoracic duct, lymphatics, and kidneys were made by Christopher Gunther Schellhammer, a professor of physic for some time in Jena in his *Introduction to Physiology*, which was published in Helmstadt in 1681, in his *Hearing*, which was published in Leyden in 1684, and in his *Dissertation on the Origin of the Lymph*,

[15] Ophthalmographia, sive oculi ejusque partum description anatomica, cui accessit nova visionis theoria. 8vo. Cantab. 1674.
[16] Anatomia chirurgica, or Istoria dell'ossa e muscoli del corpo umano, con descrizzione de'vasi. Roma, 1675.
[17] Exercitatio anatomico-medica de glandulis intestinorum. Schaffhausæ, 1677.

which was included in the *Bibliotheca Anatomica*. Blancard, who has already been acknowledged for the compilation of his useful medical lexicon, published in 1688 a textbook on anatomy containing a compilation of judicious observations made by others without acknowledgement them.

Cowper, Riely, and Boulton

The *Myotomia reformata* of William Cowper, an eminent surgeon of London, first appeared in 1694. Although this work by no means can be compared with the later works of Albinus, it far exceeded all similar works published previously, not only by its superior accuracy, but also by its descriptions of many novel muscles. The publication of Cowper's *Anatomy of the Human Body* in Oxford in 1697 initiated a controversy with Bidloo, a Dutch anatomist, when he used impressions of the plates belonging to Bidloo's great anatomical work to illustrate his own along with 40 figures engraved from his own drawings. In this splendid work, Cowper greatly improved and corrected Bidloo's descriptions of the figures in addition to adding some ingenious and valuable anatomical and surgical observations of his own. Cowper, in the course of his reply to Bidloo's accusations of plagiarism, described two new glands in the urethra, which have been termed "Cowper's mucous glands" in his honor. Cowper was also the first to describe the thoracic duct as it is found in the human subject, as anatomists before him had only described the thoracic duct as found in various animals.

In a work on the anatomy of the brain that was published in 1695 by Henry Ridley, a fellow of the College of Physicians, we find several remarks on matters that had escaped the observation of Willis and Vieussens with an account of the animal functions and muscular motion. In 1697, Richard Boulton, a practitioner of medicine and surgery in Chester, brought up in his *Treatise on the Cause of Muscular Motion* a theory of muscular motion that corresponds so closely with that of Borelli that it would almost appear to have been borrowed from it. He attempts to attribute it to a mixture of blood and nervous fluid in certain glands that he assumed to exist in the fleshy parts of the muscles. He talked about the nervous fluid as though its existence were real, not hypothetical.

Bidloo

Godfrey Bidloo was a Dutch professor of anatomy and surgery in Leyden. He published a volume of observations on amoeba found in the liver of sheep in 1698. It was his great anatomical work published in 1685 with 105 magnificent anatomical plates that resulted in the controversy with Cowper. It appears that Cowper purchased 300 impressions of Bidloo's plates for the illustration of his own work. Bidloo complained about Cowper's actions,[18] and Cowper replied to Bidloo in his *Eucharistia*, stating that he had believed the plates upon purchase to have belonged to a planned work of the deceased Swammerdam. Had this really been the case, Cowper should have stated that fact in the work and he should not have waited three years to reply to Bidloo's charge. It must be admitted that despite his service to science

[18] Gulielmus Cowper citatus coram tribunal. 4to. Lugd. Batav. 1700.

in correcting a multitude of errors and inaccuracies in Bidloo's explanations, Cowper's conduct was not justified.

Besler, Michael Campi, and Balthasar Campi

The science of botany did not advance as rapidly as might have been expected from the stimulus given to it by Gesner, Clusius, Caesalpinus, and the other great reformers of science during the preceding century. The seventeenth century hardly furnishes anything of importance. The first work that deserves to be mentioned is the splendid book *Hortus Eystensis*, which was published by Basil Besler in Nuremberg in 1612. It contains 356 atlas folio plates, on which were engraved 1,533 figures of plants at the expense of Conrad, the bishop of the diocese. However, it has many errors.

In 1623, Michael and Balthasar Campi, two eminent botanists of Lucca who were deeply read in the works of Dioscorides and those of the Arabs, compared their descriptions with the plants kept under their names in the shops, and thus detected and pointed out a multitude of errors. They also discovered many new plants during their excursions over the Alps and Appenines. In their *Nuovo Discorso*, which was published in 1623, they discussed their investigation of several of the ingredients in the composition of mithridate. In their *Specilegio Botanico*, which was published in 1650, they endeavored to prove that the *Cinnamomum* of today is altogether different from that spoken of by Dioscorides.

Brunn and Hoffman

In 1630, John James Brunn, a physician and professor of anatomy, botany, and medicine in Basil, published a pharmaceutical work.[19] An edition with notes by Gerard Blasius was published in 1680. In 1667, Caspar Hoffman published his work on officinal medicines, with a dedication to the celebrated Guy Patin. In this work, he discusses the comparative merits of law and medicine and, after quoting Aristotle, concludes that medicine is superior to law. In his preface, he disapproves of the chemical and mineral preparations used in medicine, observing that the distilled waters, from the empyreumatic impregnation that they contain, are injurious to weak stomachs; that the distilled oils exacerbate inflammation; and that spirit of wine, from whatever substance it may be obtained, injures the liver and produces dropsy and other fatal maladies in a few months. He regards antimony, arsenic, and mercury as deadly poisons, thereby suggesting that unsophisticated productions of the vegetable kingdom alone are to be relied upon.

[19] Systema Materiæ Medicæ, continens Medicamentorum simplicium et compositorum seriem ac sylvam, methodo medendi ac formulis remediorum præscribendis accommodatum. 8vo. Basil, 1630.

Charas

In 1668, Moses Charas, an eminent chemist and pharmacist, published a chemical analysis of the *Theriaca Andromachi.*[20] He discovered that its activity derived from the opiates and spices that it contained, and thus concluded that age impaired its qualities.

The following year he published results from new experiments on the viper, stating that a drop of the oil of tobacco introduced into a wound made on the viper is instantly fatal to it. He provides an anatomical description of the viper. He does not omit the venom sack in his description, but contends that its contents only become poisonous if the animal is irritated. This argument directly contradicts the experiments of Signor Redi, which demonstrate that when venom is taken from a dead viper and introduced through a quill into a wound, it is as malignant as when introduced through the fangs of a live viper. Finally, he published his work *Pharmacopée Royale, Galénique et Chymique* the same year.

Tillingius, Tournefort, and Dale

In 1679, Matthias Tillingius published his account of the plant[21] whose roots yielded the rhubarb of the shops. In the preface, he gives an account of the various productions of China. In the first part, he discusses the origin of the plant's name. In the succeeding chapters, he explains its characters, uses, and the cases to which it is applicable. In 1684, the celebrated Pitton de Tournefort published his elements of botany. However, botany continued to languish as a science until the great Linnæus kindled it into animation.

The edition of the pharmacopœia[22] published in 1689 by the College of Physicians marks the scanty improvement that had been made in this important discipline of medicine. This fact is evidenced by the complexity of the famous *Theriaca Andromachi*, which contains at least 60 ingredients. Among the contents of this pharmacopœia are the *Aqua Lumbricorum*, which is prepared from living earthworms digested in a water bath; the *Aqua omnium florum*, from cow's dung gathered in May; and the *Aqua Ranarum*, from living frogs caught in the beginning of April. The pharmacologia of Dr. Samuel Dale,[23] first published in 1693, redeems the absurdities of the pharmacopœia of the College of Physicians. It contains many useful facts on the *materia medica* and pharmacy that would not disgrace a later period and more enlightened age.

[20] Andromachus was a physician of the Roman emperor Nero. He is credited with the invention of the famous antidote for poison known as Theriaca Andromachi.

[21] Rhabarbarologia, seu curiosa Rhabarbari disquisitio, 4to. Francofurti ad Mœnum, 1679.

[22] Pharmacopœiæ Collegii Regalis Londoni remedia omnia, succincte descripta. Editio altera priori castigatior et auctior. Lond. 1689.

[23] Pharmacologia, seu Manductio ad Materiam Medicam, 4to. Londoni, 1737, Editio tertia.

Chapter 13

Medicine and Surgery in the Eighteenth Century

Boerhaave

After the death of Drelincourt, who had long served as the chair of medicine at the University of Leyden, in 1701, Herman Boerhaave (Figure 38) was nominated as his successor. He had recently become established in practice but was not yet famous. Before obtaining this appointment, Boerhaave had more leisure time than was altogether desirable. However, to a mind constituted like his, such leisure was no disadvantage. He willingly embraced the opportunity given to him to review various medical theories that had been introduced from time to time, one of which he digested more profoundly than all the rest and held an undisputed authority in the medical world for over half a century. Upon assuming the chair of medicine, Boerhaave delivered an inaugural dissertation in which he recommended the study of the works of Hippocrates, whom he admired.

After his appointment as a professor, Boerhaave quickly became famous. In 1703, he received an invitation from Groningen to accept a professorial chair at that university, an honor that he declined out of a patriotic preference for his own country and for the university that had first distinguished him. To signify their appreciation of his loyalty, the University of Leyden raised his salary. It was around this time that he published his discourse on the use of mechanical reasoning in medicine.

After the death of Peter Hotton in 1709, Boerhaave assumed the roles of curator of the university garden and professor of botany. On this occasion, Boerhaave delivered his third admirable oration,[1] in which he attributed attention to facts and observations as the best means of promoting medical knowledge in addition to the superiority of simplicity over the complex and unscientific formulae popular at the time. His celebrated *Aphorisms*,[2] an admired and universally read work, appeared the same year. His pupil for a period of nearly 20 years, Van Swieten, later published his own commentaries on this work. Boerhaave also published around the same time *Institutiones Medicæ*.[3] These works went through numerous

[1] Oratio quâ repurgatæ medicinæ facilis asseritur simplicitas.
[2] Aphorismi de cognoscendis et curandis morbis, 8vo. Lugd. Batav. 1709.
[3] Institutiones rei Medicæ in usum annuæ exercitationis domestic.

editions and were translated into almost every European language and even into Arabic. Boerhaave, now at the height of his fame, had pupils flocking to him from every part of the world. He lectured on the theories of medicine, botany, and chemistry, with a clearness and precision that surprised and delighted Haller, who attended his lectures for two years.

Figure 38. Herman Boerhaave, 1668-1738 (Image courtesy of Reynolds Historical Library, the University of Alabama at Birmingham).

Upon Bidloo's death in 1713, Boerhaave took over the professorship of the practice of medicine, and in 1718, he took over the professorship of chemistry. Thus, he was now at the head of every branch of medicine. He had so many pupils that the town of Leyden was almost unable to accommodate them. His multiple occupations obligated him to be a consultant for physicians worldwide in the event of difficult and dangerous cases. Macquer testifies that, next to Stahl, he was the most eminent chemist of his age. In 1729, as a result of deteriorating physical health, he resigned from his professorships of botany and chemistry.[4] During the year before he resigned from his professorships, he was chosen as a foreign associate of the Royal Academy of Sciences in Paris. In 1730, he was elected as a fellow of the Royal Society of London, and he was again appointed rector of the university.[5] His later years were chiefly

[4] Boerhaave delivered the following address upon resigning from his professorships in 1729, Oratio cùm cathedræ Chemiæ et Botanices valediceret. 4to. Lugduni Batavorum, 1729.

[5] Boerhaave had served in this office before in 1714. Upon stepping down, he read his discourse *De comparando certò in physicis*.

passed in domestic relaxation near Leyden. Towards the end of 1737, he developed shortness of breath and a progressively increasing sense of suffocation. He died on September 23, 1738.

Deventer

The same year that Boerhaave started his career as a professor at the University of Leyden, Henry Deventer, one of the most successful and distinguished surgeons and male midwives of his day, produced a Latin translation of a work on midwifery that he had previously published in Dutch in 1696.[6] In this work, he details the various important improvements that he had introduced to the practice of midwifery. Deventer, finding that the secret of the forceps, which was introduced by the Chamberlens, could only be purchased at a most extravagant price, and being aware of the injurious effects resulting from excessively frequent use of instruments in deliveries, decried its use. He contended that the greatest obstacle to delivery arose from the oblique position of the uterus. In all difficult cases, except when rendered impracticable by a low placed head, he passed his hand into the uterus, turned the child, and delivered it by the feet. If this maneuver was impracticable, he introduced his left hand into the posterior part of the vagina and gradually pushed back the coccyx in order to enlarge the aperture. However, if the pelvis was distorted, even this maneuver would be as ineffectual as using the Chamberlen forceps. Deventer's plan of forcing back the bone was liable to many objections, for no matter how delicately it was performed, this maneuver made the patient prone to either abscesses or incurable lameness. It would almost appear as if such accidents had occurred to Deventer himself in the course of his practice, for he later admitted the necessity of sometimes opening the head of the fetus and extracting it with a crotchet.

When an arm presented, Deventer was in the habit of passing his hand into the uterus and delivering the child by the feet in place of the cruel and unnecessary practice recommended and adopted by former writers of twisting the limb violently off. Deventer taught his pupils to distinguish between true and false labor pains and to allay the former by gently evacuating the bowels. These important improvements gave him a decided preference over Mariceau, who was almost his immediate predecessor. In 1724, he published a second part of his work on midwifery.[7] In 1733, after having been translated and published in most parts of Europe, both parts were republished together along with revisions and added material. A posthumous work of Deventer's on rickets, which Haller praises, was published in 1739.

Astruc

In 1702, John Astruc, a native of Sauve in lower Languedoc, who had just acquired the degree of bachelor of medicine in Montpellier, published a dissertation titled *De motus fermentativi causa*. It was soon followed by a variety of controversial pieces on the nature of digestion, which he accounted for with the hypothesis of fermentation excited by a peculiar

[6] Operationes Chirurgicæ, novum lumen exhibentes obstetricantibus. 4to. Lugduni Batavorum, 1701.
[7] Ulterius examen partium difficilium, Lapis Lydius obstetricum, et de necessaria cadaverum incisione. 4to. Lugduni Batavorum, 1724.

kind of leaven. This theory was contrary to the doctrine of Pitcairne and others, who attributed digestion to the mechanical action of the abdominal muscles and other muscles.

Astruc became the chair of medicine and anatomy in Toulouse in 1710. However, upon the death of Chatelain in 1716, he returned to Montpellier to take over his position. In 1720, he published a work on hydrophobia, and in 1721, he published a work on epidemics,[8] in which he asserts the contagious nature of the plague in opposition to other writers. He imagined that there was a certain analogy between the plague and syphilis. Around this time, a violent dispute arose over the question of privileges between physicians and surgeons in Paris in which Astruc participated early on. He provided evidence that it had been the former practice for surgeons to undergo an examination by physicians before they obtained a license to practice. Astruc received from Augustus II an invitation to Poland in 1729 and became his physician. However, upon finding his situation there unfavorable for his studies, he returned to France and settled in Paris, where in 1730 he received the appointment of consulting physician to the King and soon after succeeded Geoffroy as the chair of medicine at the Royal College.

In his memoirs,[9] he discusses the mineral waters of Balaruc. In 1743, he was made a member of the faculty in Paris. Although favorably received and eminently popular when they were published, his *Tractatus pathologicus* and *Tractatus therapeuticus*, which were published in 1745 and 1748, respectively, have become obsolete. His publication on syphilis[10] raised his fame to its climax and has inseparably linked his name with that disease. This work translated into all the modern languages, as it contained the most complete history, description, and treatment of syphilis. In the first part of this work, he claims that the disease was of recent origin, perfectly distinct from leprosy and every other disease, and had been introduced by the companions of Columbus from America, an opinion that had been the subject of much controversy during that period. Astruc brings forward evidence[11] to show that it was first introduced into Spain from Hispaniola and other West Indian islands towards the end of the fifteenth century. From Hispaniola, it was carried by some Spanish troops in 1495 to Naples during the war between Ferdinand of Aragon and the French, and to some of the Neapolitan women with whom both the French and the Spaniards had intercourse. In this manner, syphilis spread to both armies, and then, through commercial intercourse, gradually extended over the whole world.

In his essay on the origin of smallpox, Meade stated that syphilis was hardly known in the Russian Empire before the time of Peter the Great. However, no sooner had Peter resolved to visit other parts of Europe and sending numbers of his subjects abroad to learn trades and manufactures than these people, on their return to their own country, carried back with them this dreadful contagion.[12] For the cure of this formidable disease, Astruc relied on mercury. Shortly after the appearance of his celebrated work on syphilis, Astruc addressed to the faculty anonymously published doubts concerning smallpox inoculation.[13] In 1759, he

[8] Sur l'origine des maladies épidémiques, principalement de la peste. 1721. Disseration sur la contagion de la peste. 8vo. Toulouse, 1724.

[9] Mémoires pour l'Histoire Naturelle de Languedoc, 1737.

[10] De Morbis venereis. Paris, 1740.

[11] De Morb. vener. lib. i. cap. 10, 11.

[12] Meade's medical works, page 231.

[13] Doutes sur l'inoculation de la petite vérole, proposes à la Faculté de Paris.

published a treatise on tumors,[14] which was followed by two letters, the first of which addresses the composition of certain remedies, and the second the nature and success of certain newly proposed remedies for syphilitic affections.

Astruc is one of the first writers who exposed the popular absurdity that refers moles and other accidental marks to the force of the maternal imagination exerted during pregnancy. His work on the disorders incident to females,[15] together with that on midwifery,[16] which he did not live to complete, have both been translated into English. Astruc died on May 5, 1766, at the age of 82 years.

Arbuthnot

In 1704, a paper by the celebrated Dr. John Arbuthnot on "The constant regularity, which is observed in the births of both sexes" was read before the Royal Society, and published in their *Transactions*. In this paper, the doctor states that, despite the existence of a small excess on the part of the male births, he conceived that, from the more hazardous nature of their occupations, their numbers at an adult age were nearly equal. From this he necessarily concludes that polygamy is contrary to the law of nature and the state and deleterious to the progression of the human race, thereby making arguments in favor of polygamy, contained in Madan's celebrated *Thelypthora*, impossible to defend. Shortly after, he was called in to attend Prince George of Denmark, in consequence of the indisposition of the Prince's own attendant, Dr. Hannes. His success in this case, attracted the notice of the Queen, who then appointed him as one of her principal physicians. However, his practice does not appear to have been very extensive, since he found leisure for an excursion to Paris, upon his return from which he distinguished himself chiefly as a wit among the constellation of geniuses who adorned the Augustan reign of Queen Anne. In 1731, he published the first edition of his *Essay on the Nature of Aliments* and in the following year, his essay *On the Influence of Air on the Human Body*, both of which are founded upon the popular principles of Boerhaave. A second edition of the former of these essays appeared in 1732, with the addition of *Practical Rules of Diet in the Various Constitutions and Diseases of the Human Body*. Arbuthnot suffered from asthma for many years and died on February 27, 1735.

Mead

Richard Mead (Figure 39), the descendant of a considerable family in Buckinghamshire, was born in Stepney in August 1673. He published in 1704 his work on solar and lunar influences, founded upon principles of planetary attraction, and then recently demonstrated by the research of Sir Isaac Newton. The influence of lunar attraction in modifying the diseases to which humanity is subject is an opinion that has prevailed more or less among men in every age and every country. Mead, upon this topic, furnishes many striking proofs of the

[14] Traité des tumeurs, avec deux lettres, la lère. sur la composition de quelques remèdes, la 2de. sur la nature et les success des nouveaux remèdes qu'on propose pour la guérison des maladies vénériennes, 1759.

[15] Traité des Maladies des femmes. 12mo. 6 vols. 1761.

[16] L'art d'accoucher réduit à ses principes.

reality of this influence on disease. The writer particularly experienced this modification of the lunar influence through experiments that explored medicinal properties of the bark of the root of the Jamaican dogwood. When gathered at the full moon in April, when the tree is in flower, but before the foliage has appeared, the tincture prepared is a powerful anodyne and narcotic, whereas the tincture prepared from the bark gathered in July or August is almost inert. Mead had previously published in 1702 his mechanical account of poisons, which consists of a series of six essays. In the first of these, he shows that the ancients were fully aware of the harmless nature of reptilian poison unless introduced into the blood.

Mead had earned his degree as Doctor of Physic from the University of Padua, on December 4, 1707, and was created a Doctor of Medicine by diploma at Oxford. Mead was a strong advocate for the use of purgatives in lessening, if not preventing, the secondary fever that is often the fatal factor of smallpox. He communicated his opinions upon the subject to Drs. Radcliffe and Friend, the latter of whom adopted them and published Mead's letter on the subject, containing a number of cases illustrative of the efficacy of the practice, in his commentary on the first and third books of the epidemics of Hippocrates. Being applied to by order of the Lords of the Regency, at the time of the fatal plague in Marseilles in 1719, he inquired into the best means of preventing its introduction into England, by publishing in the following year his treatise on the plague, which was bought up with such avidity that at least seven editions were printed in one year. He died on February 16, 1754 at the age of 61 years.

Figure 39. Richard Mead, 1673-1754 (Image courtesy of Reynolds Historical Library, the University of Alabama at Birmingham).

Amand

A publication on midwifery by Pierre Amand, *Nouvelles observations sur la pratique des Accouchemens*, was printed in Paris in 1705. Amand was a successful practitioner of the obstetric art in Paris. He was born at Riez in Provence around 1650. In this publication, we meet with several cases of extra-uterine impregnation and the description of a net, contrived by the author and accompanied by a representation, for extracting the heads of fetuses when left behind after separation from the body. However ingenious this contrivance, it is now wholly superseded by the crotchet.

Haller

Albert Haller (Figure 40) was the son of an advocate of Berne in Switzerland, where he was born on October 18, 1709. He decided to join the medical profession and moved to Tubingen, where he studied anatomy under Duvernoi and Camerarius. In 1725, the reputation of the illustrious Boerhaave led him to Leyden, where, besides hearing his lectures, he dissected with Albinus, and cultivated the acquaintance of Ruysch. Returning to Tubingen, he chose *Deductu salivali Coschwiziano* as the subject of the inaugural thesis for his degree, a subject which he afterwards resumed at Leyden in 1727. After visiting England, he proceeded to Paris, where he dissected under Le Dran, but after malicious information was laid against him for having subjects in his room for dissection, he was obliged to leave and moved to Basil. In addition to studying mathematics under Bernouilli, here he acquired a taste for botany, which he had previously disliked. He tells us that in this he was inspired by the genius of the place, which the presence of the Bauhins had consecrated, and which boasted of being the residence of Stahalin.

Having returned to Berne about 1730, Haller commenced publicly lecturing on anatomy. Having acquired some celebrity abroad by the publication of various detached papers on anatomy and botany, he received an invitation in 1736 from George II, King of England, to undertake the professorship of anatomy, botany, and surgery at the recently founded University of Gottingen. Despite the unavoidable pang of separation from his native country and the connections of his wife, he did not deem prudent to reject this offer. His arrival at Gottingen was farther embittered by the death of his wife from an injury she received from the overturning of their carriage during the journey. Experiment, he was fully convinced, was alone capable of dispelling the accumulated errors of preceding centuries and purging the science of physiology from the innumerable absurdities that obscured and disgraced it. While thus engaged in the improvement of medical knowledge, he secured the friendship of Baron de Munchhausen, the Hanoverian Prime Minister, who cordially seconded and even anticipated his schemes of establishing a botanical garden, anatomical theatre, and school for surgery and midwifery for the benefit of the university. In 1738, soon after the death of his venerable preceptor Boerhaave, he promised to publish his *Prælectiones* from a copy of his own collated with others. In 1743, he commenced the publication of a series of 36 anatomical plates related chiefly to the blood vessels in situ, which were valuable additions to the study of this branch of anatomical knowledge. The first edition of his excellent *Manual of Physiology* appeared in 1747. After having filled the duties of a professor for 17 years at

Gottingen, he returned in 1753 to Berne, where he was most favorably received and rose rapidly to the highest honors of the state. While residing at Gottingen, he had a few years before been raised by the Emperor Francis and at the request of his royal patron George II to the rank of nobility by the title of Baron Von Haller.

Haller's merited celebrity procured him numerous testimonials of respect from every part of Europe, and among the rest he was requested by the King of England to accept the Chancellorship of the University of Gottingen. He died at the age of 70 years from bladder disease on December 12, 1777.

Figure 40. Albert Haller (Image courtesy of Reynolds Historical Library, the University of Alabama at Birmingham).

Anel

It had long been known that the primary, if not complete danger of sucking venom from poisoned wounds by way of the mouth, arose from the possibility of some wound, or other abrasion of the skin lining the lips and throat, absorbing the poisonous fluid into the system and ultimately into the blood. Therefore, to relieve patients of such distressing circumstances without in any manner compromising the safety of the operator had long been a medical problem of difficult solution. However, in 1707, Dominic Anel, physician to the Court of Savoy, published a paper expressly upon the solution of this problem. Among the instruments employed by Anel for the accomplishment of his benevolent design, was a siphon of fearful size. Altogether, his whole apparatus appears to have been so cumbrous and inefficient, that it seems never to have come into general use. In 1714, he also published an improvement for the treatment of aneurysms, which had been proposed by Guillemau, a pupil of Ambrose Pare. He exposed the aneurysm, passed a ligature under the artery above the seat of disease, and then, having emptied the sac, closed the wound. Upon this method Anel improved by making a longitudinal incision over the aneurysm, without wounding it, after which he made a single ligature upon the vessel close above the tumor, and left the rest to nature. A tourniquet was in all these cases applied to the limb so as to secure a free command of the artery. This treatment caused the tumor to gradually disappear by absorption. This plan experienced considerable opposition on the part of the surgeons, most of who objected to it, although it was praised by Heister and others. Heister expresses his doubt of its success in wounds of the large crural artery, so as to enable the limb to be saved. However, its perfect efficacy and safety have been fully demonstrated by more modern practitioners. However, Anel's principal work that explained his mode of treatment of fistula lacrimalis still retains its reputation, in which he describes a fine and flexible tube by which means he was able to open the lacrimal duct, wash it out with a syringe, and finally heal the passage. This excited a more intense controversy than his plan of treating aneurysms; however, it met the approbation of the College of Surgeons in Paris, and has formed the basis of all the modern improvements in the method of treating that condition.

Lommius

In 1715, three books of medical observations with a dedication by the author to the senators of that republic by Jodocus Lommius, a physician of Brussels, were printed in Amsterdam. The work contains distinct accounts of almost every malady to which the frame is subject, ending the first book with syphilis. The first book he devoted to conditions affecting the whole system; the second book, commenced with headache, addressing those which affect particular organs; and the third is reserved for the terminations and sequelae of diseases, which he discriminates into various classes, according to their severity, duration, danger, and other characteristics. A translation of these observations, preceded by a translation of Lommius's *Treatise on Continual Fevers*, was published in London in 1732 by Dr. Thomas Dale, with a dedication to the author of the pharmacologia in the last chapter.

Heister

Laurence Heister, an eminent physician, surgeon, and anatomist who was born in Frankfort on the Maine in 1683, while professor of anatomy and surgery at Altdorf in the little canton of Uri, published his *Institutions of Surgery* in German, from which it was soon translated into Latin and most of the modern languages of Europe, thereby adding much to the celebrity of its author. He also published some works on the theory and practice of medicine, founded on the mechanical doctrines of the Boerhaavian School. He also wrote a valuable practical work in quarto of highly esteemed medical, surgical, and anatomical observations.

Chicoyneau

During the dreadful plague, which depopulated the town and vicinity of Marseilles in the year 1729, Francis Chicoyneau, a native of Montpellier, where he was born in 1672 and took the degree of Doctor of Medicine in 1693, having acquired considerable celebrity both as a medical teacher and practitioner, was appointed to the medical commission sent with the task of arresting the progress of that dreadful pestilence, on the recommendation of M. Enirac, who was first physician to the Regent. In the following year, in conjunction with M.M. Verney Deidier, his colleagues in the commission, he published his remarks on this pestilence, in which he contends for its nature not being contagious and states in corroboration that he and his colleagues, acting upon this conviction, entered the chambers of the sick without fear. Afterwards, without suffering from infection, they were led to conclude that it was not contagious, an opinion which allowed mischief of the inhabitants by leading them to neglect the proper precautions and thus contribute to the diffusion of the evil. After being ordered by the king to collect the opinions of different physicians on the subject of the plague, in particular, all the facts and observations which had been made at Marseilles, he prepared and published a quarto volume containing the result of his inquiries, drawn up with great candor and highly valuable from the collection of facts that it contained. Chicoyneau succeeded his father-in-law, Chirac, upon his death in 1731, as first physician to the King, and was also made counselor of state and honorary member of the Academy of Sciences. He died in 1752 at the age of 80 years.

Cheyne

George Cheyne, a native of Scotland, was born in 1670, and in 1722 published the first edition of an essay on gout originally designed for the private use of a gentleman of the name of Tennison, but was, as he informs us, merely an abstract of a larger work that he had not leisure to complete. In this work, the theory of gout is founded entirely upon that mechanical hypothesis that was the fashionable doctrine of the day, and he explains that the reason some persons fall victim to this dreadful disorder, while others with constitutions apparently similar in every respect, are exempt, lies within the difference in the capacity of the vessels. He strongly advocates the use of the bath waters, from which he had derived great benefit, both in this and other chronic disorders. In 1724 he published an essay dedicated to Sir Joseph Jekyl, Master of the Rolls, who had been a patient of his. In this work, which appears to have suggested a similar one by an eminent Scotch baronet in our own days, he points out the necessity of the strictest attention both to diet and regimen for the preservation or restoration of health. His own case furnished a striking and satisfactory illustration of the recommended plan's efficacy, for, having been a naturally indulgent person, his corpulence had increased to such a degree that by the time he had reached the meridian of life he found himself so unwieldy and lethargic and embarrassed by such difficulty of breathing that he became alarmed. He was determined upon adopting a total change of life, and confined himself to a milk and vegetable diet with a total abstinence from fermented liquors. By a rigid adherence to this plan, he soon found himself relieved from the most distressing symptoms under which he suffered, and was so strongly impressed with the beneficial effects of the method he pursued that he was induced to make it public for the advantage of others similarly circumstanced. In 1724 appeared a fourth edition of his work on fevers, with an essay on the improvements of the theory of medicine, the whole founded upon the mechanical doctrines of the day. The work by which perhaps his name is best known is his *English Malady*, which first appeared in 1733 and acquired so much popularity as to reach to a fourth edition in the following year. He died in Bath in 1742 at the age of 72 years.

Van Swieten

Amongst the most distinguished and favorite pupils of the illustrious Boerhaave, we find Gerard Van Swieten, the descendant of an ancient and respectable family in the Low Countries, who, after a preliminary course of philosophy at Louvain, where he preeminently distinguished himself by his industry and talents, moved to Leyden, where he studied medicine under the great Boerhaave. After seven years, he earned his degree of Doctor of Medicine, in 1725. Having previously attained a distinguished rank among men of science, he was almost immediately appointed to a medical professorship, which he held with distinguished reputation and ability for many years. His lectures, similar to those of his colleague and instructor Boerhaave, attracted an immense concourse of pupils from every part of the world. His eminent success and growing popularity eventually roused the envy and excited the malice of those who felt mortified by the consciousness of their own comparative inferiority, and wished for his removal out of their own insecurity. They referenced an obsolete and almost forgotten provision of the laws of Holland prohibiting those who professed a faith differing from the established religion of the state from holding any public appointment. Because of this, Van Swieten was obliged to resign a professorship, which he had held with equal honor to himself and advantage to the university and employed his leisure in writing commentaries on Boerhaave's aphorisms, the first volume of which had appeared, and the second was nearly ready, when his reputation procured him an invitation from the court of Vienna. This he accepted, and moved there in 1745, after stipulating for permission to follow his usual course of life. He eventually came to lead the medical school in Vienna. Through his influence with the Empress, the college was rebuilt, with the addition of a chemical laboratory, schools of anatomy and surgery, a clinical establishment at one of the hospitals, and the foundation of a botanical garden. He published a work on the diseases of armies, which possesses considerable merit. It is his commentaries on the aphorisms of Boerhaave that will hand his name down to posterity. Among other observations, he remarks, that people who by misfortune have been reduced from affluence to the necessity of earning their bread by bodily labor and exchanging a luxurious table and indolent life for spare diet and active exertion have become permanently cured of gout. After enjoying unbroken health for a long succession of years, his constitution eventually began to fail, and, after a progressive decline for three years, he was attacked with gangrene in one of his toes, which proved fatal. He died in Schoenbrun in 1772 at the age of 73 years.

Freind

In 1725, John Freind (Figure 41), a physician of great eminence and extensive erudition, published the first volume of his *History of Physic.*[17] A second volume of the work appeared in the following year. In the first of these volumes, he acknowledges the Greek writers who flourished since the days of Galen, noting with precision the era in which Oribasius, Etius, Alexander, and Paulus flourished, which former historians, even those of the best repute, had left confused and unsatisfactory. He has also preserved many valuable fragments of authors who wrote before the days of Galen, pointed out various improvements in surgery, and allotted to the ancients many facts and discoveries falsely claimed by the moderns, such as the knowledge of the salivary glands. He also dwells upon those uncommon conditions that have been noticed by the ancients, such as Guinea worm and lycanthropy. He commends ancient methods of cure. He makes many useful and curious observations on aneurysm, inguinal hernia, the discovery of the circulation, and other matters equally interesting and important. In the second volume, he shows the origin and progress of literature among the Arabs and the zeal and munificence of the Caliphs in promoting improvement, determines the eras of the Arabian physicians, and shows the portion of improvement that medicine owes to each. He addresses maladies either known to the Arabian writers alone or first mentioned by them, such as smallpox, which was utterly unknown to the Greek writers. He points out the improvements made by them in practice by the substitution of milder purgatives for those in use before, traces the origin of chemistry to them, and introduces preparations still retained in our pharmacopeias. Leaving the Arabs, he goes on to the later ages, to the revival of medicine in Europe, and to the schools of Salernum, Naples, Montpellier, and Bologna. The introduction of chemistry he ascribes chiefly to Bacon, whose various discoveries in optics, catoptrics, dioptrics, mathematics, astronomy, and natural philosophy he briefly notices. He gives an account of the pestilence that raged so extensively in 1348 and describes the novel disorders that appeared in the following century, such as the sweating sickness, which made its first appearance in 1483; scurvy, which attacked the Portuguese explorers; and syphilis, which was the price paid by Europe for the discovery of America. It was translated into Latin by Dr. John Wigan and forms part of the edition of his works published at Paris in 1735.[18]

Freind was born in 1675 in Croughton, Northampton, of which his father was rector, and received his education at Christ Church, Oxford, when the celebrated Dr. Aldrich was dean. Upon earning his master's degree on April 12, 1701, and his bachelor's degree in medicine on June 1, 1703, he was admitted to his doctor's degree by diploma on July 12, 1707. In the following year, he was appointed lecturer in chemistry at the university. In 1705, he accompanied the Earl of Peterborough as physician to the forces into Spain. Upon his return, he wrote a narrative of his experiences there. In 1709, he published his lectures on chemistry, in which he accounted for almost all the phenomena regarding the principles of the Newtonian philosophy. In 1712, he was chosen as a fellow of the Royal Society, an honor that he had earned by his services to science. In the same year, he accompanied the Duke of Ormond to Flanders, bringing back copious additions to his stores of knowledge. In 1716, he

[17] The history of physic, from the time of Galen to the beginning of the sixteenth century, 2 vols. 8vo. Lond. 1725-6.

[18] Johannis Freind MD., Serenissimæ Reginæ Carolinæ Archiatri, Opera Omnia Medica. Editio altera, Londinensi multo correctior et accuratior, 4to. Parisiis, 1735.

published the first and third books of the epidemics of Hippocrates, which were accompanied by nine commentaries on fevers. In 1719, he published a letter to Dr. Mead on the use of purgatives in the secondary fever of confluent smallpox, wherein he treats the subject with his usual skill and perspicuity. In 1720, he delivered the Harveyan Oration. Freind died at the age of 52 years in 1728.

Figure 41. John Freind (1675-1728) from *The history of physick, from the time of Galen, to the beginning of the sixteenth century*. London, Printed for M. Cooper, at the Globe in Pater Noster Row, MDCCL [1750].

Douglas

Dr. James Douglas, a distinguished anatomy teacher in his time and the immediate predecessor of the celebrated Dr. John Hunter, published in 1726 *A history of the lateral operation for the stone*. He published a second edition of this work in 1733, which was accompanied by an appendix exhibiting a comparative view of the method of operating employed by different lithotomists, and more especially that recommended by Cheselden. His brother John, who was surgeon to the Westminster infirmary and author of several controversial works, also wrote an account of the high operation for the stone, which he practiced.

Barry

In 1726, Dr. Edward Barry, a native of Dublin who studied and earned his doctor's degree at Leyden in 1719, published a treatise[19] on consumption of the lungs, in the preface to which he adds a concise review of the progress of medicine from the time of Hippocrates to that of Harvey. In this review, he demonstrates the small improvement that was really made in all that time. He gives particular credit to Aretæus for having first orderly arranged the inaccurate teachings of Hippocrates, and points out the service performed by Galen in collecting the scattered diagnostics and prognostics dispersed throughout the writings of former physicians, but censures him at the same time for his introducing the peripatetic philosophy into the theory of disease, thus not only retarding the improvement of physic, but also discrediting it with fiction and errors. From the time of Galen to that of Paracelsus, writers did little more than transcribe Galen and Aristotle, taking little notice of Hippocrates until Hollerius and Dinetus revived his doctrines in France. However, Paracelsus violently overturned the Galenic doctrines and substituted them with chemistry, explaining all the operations of the animal frame with the theory of fermentations.

Hancocke

About this time, the use of cold water began to come into vogue, and a work in favor of this practice titled *Febrifugum Magnum* was published by Dr. Hancocke. Also, an anonymous pamphlet[20] of nearly 300 pages appeared in 1726 that recommended an extension of Dr. Hancocke's plan to phrenitis, melancholy, apoplexy, palsy, catarrh, and other complaints. In phrenitis, the author recommends confining the patient to cold water almost entirely for his consumption, washing the head frequently with cold water, and using the cold bath freely. He recommends a trial of the diaphoretic properties of cold water, advising that for this purpose, after the use of the cold bath, the patient should take a large sip of cold water and be immediately put to bed and warmly covered up. The reason he assigns for giving cold water immediately on coming out of the cold bath is "because we find all people after bathing, as soon as they put on their clothes, are inclined to a breathing sweat." The internal use of cold

[19] A treatise on consumption of the lungs, with a previous account of nutrition, and of the structure and use of the lungs. 8vo. Dublin, 1726. A reprint appeared in London in the following year, and in 1759 he published a fourth edition greatly enlarged and improved, under the title of *A treatise on the three digestions and discharges of the human body, and the diseases of their principal organs*.

[20] Febrifugum Magnum Morbifugum Magnum, or the grand febrifuge improved. Being an essay to make it probable that common water is good for many distempers that are not mentioned in Dr. Hancocke's Febrifugum Magnum. 8vo. Lond. 1726.

water in ardent fevers had been recommended by ancient practitioners, such as Hippocrates, Galen, and Celsus, and modern practitioners, such as Cardan and Hoffman.[21]

Desault

In 1727, a small work[22] designed as a kind of domestic medicine was published in Paris by Peter Desault, a native and graduate of Bourdeaux who also authored a work on gout and syphilis, which last he professed to cure without salivation. Desault's practice excited some opposition, which called forth a vindication in a work[23] that he published in 1736 on the subject of calculus, wherein he strongly objects to the operation of lithotomy, regarding it as perfectly unnecessary since the stone dissolved, without the aid of cutting, by drinking the waters of Bareges and injecting them into the bladder. These waters had acquired a high reputation for their supposed lithontriptic powers and were believed to be as potent in relieving the pain and dissolving calculi in the kidneys, ureters, or bladder as Mrs. Stephen's celebrated remedy of lime water or Castile soap, both of which came into vogue about the same time. They are still employed to alleviate pain but are no longer regarded as possessing lithontriptic powers. In the second volume of this work, Desault addresses hydrophobia and expresses his belief that this formidable malady could be treated by mercurial friction; however, this opinion is not corroborated by experience.

Chamberlen's Obstetric Forceps Revisited

The inventors of the obstetric forceps, which was originally invented in the seventeeth century, only revealed its nature or use to persons willing to purchase the knowledge at an extravagant charge, and purposely threw a veil of mystery over their invention. However, in 1732, the secret was first partially revealed by Mr. Butter, who published an account of its construction and use in the third volume of *Medical Essays*, and more fully by Edmond Chapman, a distinguished surgeon and accoucheur who has the merit of having been the first to give a representation and description[24] of it in 1732, more than 60 years after its first introduction by its inventors. Chapman distinguished himself in London by his skill and experience in difficult labors. The Chamberlens had guarded the secret of their instrument so carefully that no account of it had been published before the descriptions by Butter and Chapman were written. Chapman's publication was a work of considerable merit, containing an accurate figure and description of this instrument, which he had improved considerably, and accompanied by an ample detail of cases.

---

[21] Hoffmanni Opera, vol. 1. p. 479

[22] Nouvelles découvertes concernant la santé, et les maladies les plus fréquentes. Desault, 12mo. Paris, 1727.

[23] Dissertation sur la Pierre des reins et de la vessie, 3 vols. 12mo. Paris, 1736.

[24] Treatise on the improvement of midwifery, chiefly in regard to the operation, with cases, 8vo. Lond. 1732.

Frère Jean de St. Cosme

Frère Jean de St. Cosme, or Come, a monk of the order of Feuillans, acquired great reputation by his skill in lithotomy, which he practiced without compensation from the most impartial motives of humanity. He had been regularly bred a surgeon himself before he retired from the world. He operated with what he called his *lithotome caché*, a hollow tube concealing a knife, with which he cut through the prostate into the bladder and thereby made the aperture sufficiently large to extract the stone without injury to the parts, a result to which he probably owed his great success. The surgeons, alarmed at his growing celebrity, applied to the King to forbid his practicing, but failed in this goal. Thus, they strove to check him through the medium of the press, criticizing his dissertation on the mode of operating, which was published in the *Journal des Savans* in 1748, and attacked in the following year by M. de Cat,[25] to which attack Come replied in 1751 in a work[26] consisting of a number of cases and observations, wherein he admits some instances of failure and the loss of one patient by hemorrhage, but also challenges his adversaries to produce as copious lists of successful cases as his. It is probable that his success arose more from his own personal dexterity in operating than from any superiority in the instrument he employed.

Van Doevern

Walter van Doevern, a skillful anatomist, accoucheur, and professor of medicine in Leyden, published in 1753 a treatise[27] on intestinal worms containing an accurate description of the various kinds that infest the human body. His most valuable work[28] is his anatomical account in which he notices the rupture of the bladder in pregnant women and of the uterus during a difficult labor. He also speaks of the successful extirpation of a polyp in the uterus by means of a ligature.

Huxham

John Huxham was an eminent practitioner in Plymouth, where he enjoyed a considerable reputation and, if we may judge from his writings, an extensive practice. He distinguished himself considerably as a medical writer about this period, having, as he informs us in the preface to his essay on fevers,[29] of which the third edition appeared in 1757, published about 10 years before a volume of observations on air and epidemic diseases from 1727 to the end of 1737,[30] and having, at the time of writing that preface, an additional volume of similar observations from 1738 to 1747 inclusive ready for the press. Huxham, although he founded

[25] Lettre au sujet de lithotome cache, &c. contre F. Cosme, dissertation. Par. 1749.

[26] Recueil des pieces importantes sur l'opération de la Taille. 1751.

[27] Dissertatio de vermibus in intestinis hominum genitis. Lugd. Bat. 1753.

[28] Observationum anatomicarum ad monstrorum historiam, anatomiam, pathologiam, et artem obstetricam præcipue spectantium, 4to. Groningen, 1765.

[29] An Essay on Fevers, to which is now added a dissertation on the malignant ulcerous sore throat. Third edition, Lond. 1757, 8vo.

[30] Observationes de Aëre et Morbis Epidemicis, &c. 3v. 8vo.

his practice on the basis of experience, was an advocate for the study of the principles and practice of the ancients, especially of Hippocrates. Although he does not imply that a man cannot become a good physician without being acquainted with the ancients, he expresses his firm conviction that he will make a better physician by this means. He believes that had the course laid down by Hippocrates been strictly followed by all of his successors, the art of healing would have been carried to far greater perfection than it has been. He remarks that none has followed Hippocrates more closely than Aretæus of Cappadocia, who has even affected his very words and style. He surprisingly expresses that no one should have noticed Aretæus before Ætius, who wrote in the fifth century, and declares his conviction that Aretæus did not flourish until after the time of Nero. He notes the many coincidences between the writings of Archigenes and Aretæus, especially their both recommending some particular remedies scarcely to be met with in the works of other writers, such as the external use of cantharides, which is only mentioned by these writers and Celsus. He praises Cælius Aurelianus for the accuracy in his descriptions of disease. In his chapter on putrid and malignant fevers, he points out an error committed by the great Sydenham in regarding all fevers as inflammatory, and shows that however applicable the depletory practice Sydenham recommends may be to those of a truly inflammatory character, it is perfectly unsuitable for those of the typhoid character.

Of this celebrated practitioner but few biographical notices exist, and these are chiefly to be found in an account of his life written, from very meager materials, by the late Dr. Woollcombe of Plymouth, who has done all the justice in his power to the memory of his fellow townsman. We learn from his works that Huxham was a fellow of the College of Physicians. From other sources, we learn that he was father to John Corham Huxham, of Exeter College, Oxford, who took the degree of master on March 20, 1746, and is the only person of that name in the catalogue of graduates published in 1801. Dr. Huxham was a fellow of the Royal Society, to whose transactions he contributed many papers on pathology and morbid anatomy. From the short sketch inserted in Rees's *Cyclopedia*, he appears to have died at Plymouth in 1768, and it is to be regretted that his son, who published in 1770 the third volume of his observations on the epidemics prevalent in the neighborhood of Plymouth, should not have availed himself of that opportunity to perpetuate some biographical particulars of a man who deserved so well of science.

The Medical School of Philadelphia

For the first time in the annals of medicine, our attention is demanded to the medical transactions of our brethren across the Atlantic Ocean, among whom the first dawning of a medical school appears in the courses of lectures on anatomy and surgery delivered in 1763 and 1764 to a small class at Philadelphia by Dr. William Shippen. In 1765, lectures on the institutes of medicine and the practice of physic were read at the same place by Dr. John Morgan, and again in 1768 by the venerable Dr. Kuhn, who, having been a pupil of Linnaeus and graduated the preceding year from Edinburgh, was made professor of botany and materia medica, to which catalog of teachers the name of the illustrious Benjamin Rush, the "Hippocrates of Colombia", was added upon his return to America in 1769, after having completed his studies at Edinburgh under medical luminaries such as the elder Monro, Gregory, Cullen, and Black.

However, the organization of the medical school at Philadelphia cannot be regarded as complete, or its existence established upon a permanent basis, before 1791, in which the Medical College at Philadelphia was incorporated into the University of Pennsylvania and Dr. Rush was appointed professor of the institutes of medicine and clinical practice. This eminent professor and distinguished writer, who may justly be regarded as the founder of medical science in his native country, terminated his active life on April 18, 1814, at the age of 69 years. His descriptions of disease are exceeded by none in detail and accuracy and are excellent models for succeeding writers. His accounts of the several epidemics of the United States rank with those of Sydenham and Boerhaave.

Heberden

To the suggestions of the learned and classical Heberden given to the College of Physicians in 1766, science is indebted for the publication of the *Medical Transactions*, in which he proposed to embody such observations as might have occurred to any of the members of that learned body as being calculated to illustrate the history or cure of diseases. He was the first to describe angina pectoris as an idiopathic disease. He was the first to accurately describe chicken pox and clearly distinguish chickenpox from smallpox, with which it was frequently confused. After his death in 1801 at the age of 91 years, his commentaries on the history and cure of diseases[31] were first published in 1802 by his son, Dr. William Heberden, with a dedication prefixed to George III and a concise biographical sketch of his father.

Sultzer and Galvanism

The first printed notice we have of the phenomena of galvanism is in a work titled *The General Theory of Pleasures*, which was published in 1767 by George Sultzer, a native of Winterther in Zurich. In this work, he mentions the sensation produced upon the tongue when the tongue is placed between two different metals in contact with each other. All natural phenomena at this time were attributed to vibration, and Sultzer based his explanation of the sensation thus produced on this principle. This solution was satisfying at the time, and the brilliant discovery of galvanism was forced to be hidden for nearly 40 years until it was brought to light by the genius of Galvani and illustrated by the research of succeeding philosophers.

[31] Gulielmi Heberden Commentarii de Morborum Historia et Curatione, 1 vol. 8vo. 1802. Editio altera, 1807.

Buchan

The first edition of William Buchan's celebrated work *Domestic Medicine*, founded upon the model of Tissot's *Avis aux peuples*, to which it is in no respect inferior, appeared in 1771. Whatever its merits, this work acquired popularity almost unprecedented in the annals of bibliography from the instant of its publication. This plan of exhibiting the method of treating diseases in a popular form has been strongly, but unjustly, criticized as tending to lower the esteem of the profession in public estimation and to diminish their gains.

Jenner

A somewhat remarkable circumstance occurred to the celebrated Dr. Jenner, an eminent practitioner in the county of Gloucester, that there were many of those whom he met within practice whom seemed to be immune to smallpox, and so around 1776, he was led to undertake the investigation of the cause of so remarkable a fact. In the process, he found that those non-susceptible to the contagion of variola had been infected with the casual cow pox, an ailment familiar from time immemorial in the dairy districts of that county, and believed by the inhabitants to prevent the smallpox. Upon further inquiry, he found some apparent exceptions to this rule, but, pushing his investigations further, he discovered these exceptions to result in circumstances of the cow being subject to a variety of pustular eruptions differing widely in their nature, but indiscriminately classified by the ignorant under the name of "cow pox". After conducting a variety of research, it eventually occurred to him that this disease, like variola, might be communicated by inoculation. In 1778, Mr. Cline successfully inoculated a child with virus received from Dr. Jenner, who had in the preceding month given his first publication on the subject to the world, and then exposed the child to the test of variolous inoculation in three places without producing the slightest effect. Upon this occasion, Mr. Cline assured Dr. Jenner that both he and Dr. Lister, who had been physician to the smallpox hospital, were fully convinced of the efficacy of the vaccine inoculation, and that the substitution of this mild, manageable, uncontagious disease for smallpox promised to be one of the greatest improvements in modern practice. Thus, the merit indisputably belongs to Mr. Cline of having vaccinated the first patient in London, but has been attempted to be taken from him by the claim of another practitioner, which considerably lacks in similar foundation.

Cullen and Gregory

During the greater part of this century, the medical school at Edinburgh, of which the elder Monro may justly be regarded as the founder, enjoyed a succession of teachers unrivaled perhaps by any university in the world. Among these teachers is William Cullen, who was born in Lanark in 1712. After experiencing a variety of calamities, he was appointed professor of chemistry at Glasgow in 1746. He was advanced to the chair of medicine in 1751. He was invited to fill the chemical chair at Edinburgh when Dr. Plummer died, and he moved there in 1756. His colleague in anatomy was the illustrious Monro. Upon the death of Dr. Alsten in 1760, Cullen succeeded to the chair of materia medica. Upon taking possession of the chair of medicine at Edinburgh, Cullen set about to amend the defects in the system of Boerhaave, which he had learned in his youth and was the prevailing and indeed only system in existence. He continued to deliver his lectures until a few months before his death. He died on February 5, 1790, at the age of 77 years. Cullen was appointed, in conjunction with James Gregory, to replace Dr. Rutherford as the chair of medicine in 1776, but the unfortunate death of his valuable colleague the following year left him as the sole chair of medicine. Gregory's work on the theory of medicine has conferred immortality upon his name and honor on his country.

Discovery of Galvanism

In an Italian work published in 1786, five years before the appearance of Galvani's account of his discoveries, we find an additional circumstance connected with the history of those discoveries. The narrative is given by Cottugno. "A medical student feeling a smarting sensation in the lower extremity of his leg, applied his hand to the part, and caught a mouse by which he had been bitten. Having killed it, he decided to dissect it and upon touching the intercostal nerve with his knife, was not a little surprised at experiencing an electrical sensation, sufficiently powerful to benumb his hand."

This fact awakened the curiosity of Vassali of the Royal Academy of Turin, who made in response a series of experiments on the subject, the details of which appeared in 1789. Still, however, the experiment related by Sultzer, the fact communicated by Cottugno, and the xperiments conducted by Vassali continued only as isolated particulars, unconnected by any general or comprehensive views and forming only the useless and disjointed fragments of a science from which no rational principles could be deduced or practical utility derived, until 1792, when the celebrated Galvani laid the foundation of that splendid science, which remained for the distinguished Englishman Davy to complete, by the publication of his interesting experiments in the first work that he published on the subject. This publication immediately excited universal attention.

Louis Galvani (Figure 42) was born in Bologna on September 9, 1737. One of Galvani's assistants accidentally touched the nerve of the thigh of a frog with the point of a knife, and the muscles of the limbs became instantly and powerfully convulsed. Madame Galvani, who was present, was shocked by the singularity of the phenomenon and imagined at the instant of every spark passing from the conductor. Relaying this observation to her husband, he determined to investigate the fact. Upon bringing the point of the scalpel, which he held in his hand, in contact with the crural nerves of one of the frogs, he found, as his wife had pointed out, that the muscular contractions were renewed as often as a spark was taken from the conductor. As this might have resulted simply from the irritation of the scalpel and not from the disengagement of the spark, he then decided to touch the same nerves while the electrical machine continued in a state of quiescence. This time, he did not excite the slightest commotion. Upon this foundation, he constructed the theory of the science that bears his name.

Figure 42. Louis Galvani, 1737-1798 (Image courtesy of Reynolds Historical Library, the University of Alabama at Birmingham).

Chapter 14

Anatomy and Physiology in the Eighteenth Century

Douglas

Among the earliest and most successful anatomists during the eighteenth century was James Douglas, the predecessor of the distinguished William Hunter (Figure 43). Douglas was born in Scotland in 1675. After he completed his studies, he settled as a teacher in London, where he experienced great success. Haller, who visited him, speaks highly both of him and his preparations, which were made with much skill and calculated to show both the motions of the joints and the internal structure of the bones. At that time, he appears to have been planning an extensive work on anatomy, which he did not complete. Upon William Hunter's first arrival in London, Douglas made him his assistant for dissections and gave him an opportunity to improve himself by attending St. George's Hospital.[1] In 1707, Douglas published *Myographia Comparata Specimen*,[2] a work containing the most accurate account of the muscles that had yet appeared and comparing the muscles in a man and in a dog. This work contains accurate accounts of the muscles that had yet to appear. Douglas's anatomical bibliography appeared in 1715. This work contains a tolerably correct account of several works on anatomy with biographical sketches of their respective authors,[3] and an improved edition appeared in Leyden in 1734. His next publication was on the subject of the peritoneum and the cellular membrane that is situated outside it.[4] He had a short time before published a work on lithotomy. In addition, he contributed many papers to the Royal Society on the anatomy of the uterus and neighboring vessels with a variety of surgical cases that were published in the *Philosophical Transactions*. Douglas died in 1742.

---

[1] Douglas's daughter, Martha Jane, married William Hunter. However, she died at the young age of 28 years. See, Bryn Thomas K. James Douglas of the pouch, 1675-1742. Br Med J 1960; 1, 1649-50.

[2] Myographiæ comparatæ specimen, 12mo. Lond. 1707.

[3] Bibliographiæ anatomicæ specimen, seu catalogus pene omnium auctorum qui ab Hippocrate ad Harveium rem anatomicam illustrarunt, 8vo. Lond. 1715.

[4] A description of the peritoneum, and that part of the membrana cellucaris which lies on the outside of it, 4to. Lond. 1730.

Figure 43. William Hunter, 1718-1783 (Image courtesy of Reynolds Historical Library, the University of Alabama at Birmingham).

Cheselden

William Cheselden distinguished himself at an early age through his proficiency in anatomy. He studied anatomy under Cowper. In 1711, at the age of 22 years, he became a public lecturer on anatomy and surgery. In 1713, he published *Anatomical Description of the Human Body*,[5] which was accompanied by some select surgical cases and a syllabus of his lectures. This work acquired such popularity that it passed rapidly through six editions. To the fourth edition and all the editions thereafter, Cheselden added a short account of the operation of lithotomy in an appendix, which he successfully performed on nine patients at St.

[5] Anatomical description of the human body, with plates, 8vo. Lond. 1713.

Thomas's Hospital in the manner recommended by his contemporary James Douglas. However, the procedure not being successful in later trials, he adopted the method recommended by the Dutch physician Johann Rau, in which he made such improvements that he successfully operated on the first 20 patients without any failures. Notwithstanding the candor with which he admitted Douglas's improvements concerning this operation, Cheselden was criticized in an anonymous pamphlet titled *Lithotomuscastratus*, which was attributed to some of the supporters of Douglas.

Heister

Laurence Heister was a celebrated physician, surgeon, anatomist, and botanist. His most distinguished work was *Compendium Anatomicum*,[6] which was first published in 1717. This work, valuable for its conciseness and clearness as a textbook on physiology and anatomy, became a very popular book after going through multiple editions. It superseded the work of Peter Verheyen, the defects of which Heister clearly exposed in his preface to *Compendium Anatomicum*, and which he reasonably criticized for omitting some of the principal discoveries of English and Italian anatomists and anatomists of other nationalities.

Monro and the Medical School of Edinburgh

A very distinguished period in the history of British medicine is hallmarked by the founding of the medical school in Edinburgh, which attained the first rank among the medical seminaries of the world. The establishment of this school is chiefly credited to Alexander Monro.[7] He was born in September 1697 in London. His father was a surgeon in King William's army. Upon leaving the army, Mr. Monro settled in Edinburgh. Perceiving his son's early indications for being a genius and having a strong inclination towards medicine, he carefully oversaw his early education and then sent him to London to study anatomy under Cheselden. The young Monro studied anatomy with the greatest diligence and devoted himself with the utmost assiduity to the dissection and preparation of anatomical specimens. While in London, he laid the foundation for his most important work on the bones, *Osteology, or Treatise on the Anatomy of the Bones.*[8] From London, he proceeded to Paris, and then to Leyden in the autumn of 1718. In Leyden, he studied under Boerhaave. Returning to Edinburgh in 1719, Monro (Figure 44) was appointed professor and demonstrator of anatomy to the company of surgeons and soon after started lecturing, employing the preparations he had made during his anatomical studies as illustrations. At the same time, Alston, to whom Cullen succeeded as the chair of *materia medica* in 1760, started a course of lectures on *materia medica* and botany. These lectures were the first public medical lectures delivered in

[6] Compendium Anatomicum, 8vo. 1717.

[7] Monro's family was notorious for its contribution to anatomy and medicine. Alexander Monro primus (1697-1767), secundus (1733-1817) and tertius (1773-1859) held the chair of anatomy at the school of medicine in Edinburgh for over 120 years. See, Wu OC, et al. The remarkable medical lineage of the Monro family, contributions of Alexander primus, secundus, and tertius. J Neurosurg. 2012; 116, 1337-46.

[8] Osteology, or Treatise on the Anatomy of the Bones. Edinb. 1726.

Edinburgh, and were the basis of the school of medicine. The plan for the school's establishment originated with Monro's father, who invited the whole college of physicians and the company of surgeons without previously acquainting his son with them. Being confronted with an entirely different audience than what he expected, Monro was compelled to deliver lectures extemporaneously and not resort to the notes he had prepared beforehand. Through the indefatigable efforts of Monro's father, a regular series of medical lectures was established in Edinburgh. An infirmary was erected, endowed, and incorporated by charter, following which a course of clinical lectures on surgery was started by Monro. In 1748, Rutherford stated a course of clinical lectures on medical cases.

Monro's *Osteology*, which he started to develop when only a student in London, was now completed and published in Edinburgh in 1726. It became very popular and passed through numerous editions and was translated into many European languages. Monro added to later additions a concise account of the nerves, lacteal system, and thoracic duct. However, Monro was not content with the services he had performed for science, and his zeal led to the formation of a society for the publication of papers on professional subjects, to which he himself contributed enormously. In 1759, he relinquished the chair of anatomy, which he had held for nearly 40 years, to his son. However, he still continued to deliver clinical lectures at the infirmary. He died from an ulcer secondary to a fungal infection of the bladder and rectum at the age of 70 years on July 10, 1767.

Figure 44. Alexander Monro primus, 1697-1767 (Image courtesy of Reynolds Historical Library, the University of Alabama at Birmingham).

Paccioni

Antonio Paccioni, a native of Reggio, was distinguished for his anatomical research, especially that on the brain and dura mater. In 1701, he published his first work on the structure of the dura mater.[9] He maintained that the dura mater has a muscular and contractile nature and claimed that it acts through alternate compression, by means of its connection with the tentorium, upon the cerebrum and cerebellum. After Baglivi, Fautonie, and others contradicted this opinion, he published a second dissertation on the subject, in which he announced the discovery of glands inside the longitudinal sinus and connected with the openings of the veins, in 1705.[10] These glands are now referred to as Pacchionian bodies or granulations. Although their existence has been acknowledged, the nature of their function remains unknown. This discovery initiated disputes with other anatomists and led Paccioni to produce vindications of his opinions and discoveries, in which he defended his doctrine of the glandular structure of those parts.[11] In 1721, Paccioni published his final dissertation on the dura matter.[12]

Winslow

In 1723, James Benignus Winslow, an eminent professor of anatomy, medicine, and surgery at the University of Paris, published *Exposition Anatomique de la structure du corps humain*,[13] which was popular for its clearness, conciseness, and organization.

Swedenborg

Emanuel Swedenborg was distinguished for his talents and erudition in his early life. Besides other works not related to medicine, he published his great work on anatomy and physiology in 1742.[14] In the first chapter of this work, he discusses the composition and what he terms the "genuine essence" of the blood. In the second chapter, he discusses the circulation, arteries, and veins. In the third chapter, he discusses the formation of the chick in the egg and the first rudiments, or *inchoamenta*, of the arteries, veins, and heart. In the fourth chapter, he discusses the circulation before birth, foramen ovale, and arterial canal of the fetal heart. In the fifth chapter, he discusses the heart of the sea turtle. In the sixth chapter, he discusses the coronary vessels and other vessels belonging to the heart. In the seventh chapter, he discusses the motion of the heart in adults, which he illustrates with a plate. In the eighth

---

[9] De Duræ Matris fabrica et usu, disquisitio anatomica. Rome, 1701.

[10] Dissertatio epistolaris de glandulis conglobates Duræ Meningis humanæ, indeque ortis Lymphaticis, ad piam matrem productis, ad clarissimum virum Lucam Schroeckium. Rome, 1705.

[11] Dissertationes binæ ad spectatissimum virum Johannem Fantonum datæ, &c. Rome, 1713.

[12] Dissertationes physic-anatomicæ de Dura Meninge humana, novis experimentis et lucubrationibus auctæ et illustratæ. Rome, 1721.

[13] Exposition Anatomique de la structure du corps humain. 4to. 1723.

[14] Œconomia Regni Animalis in transactiones divisa, quarum hæc prima de sanguine ejus arteriis venis et corde agit. 4to. Amstelodami. 1742.

and final chapter, he gives an introduction to a rational physiology, or that science that treats of the essence and nature of the soul. Swedenborg developed an idea of somatotopic organization of the cerebral cortex, described the role of the pituitary gland in maintaining normal function, and discovered the cerebral perivascular spaces and the foramen of Magendie.[15]

Theophilus de Bordeau

Theophilus de Bordeau was a French physician who lived from 1722 to 1776. He published a small treatise on the pulse titled *Recherches sur le pouls par rapport aux crises* in 1756.[16] He went far beyond Solano and what can be followed in practice in his discrimination of the endless variations in the pulse. In another work, which addresses the mucous tissue and cellular membrane, he disingenuously claimed some properties of the cellular membrane as his own, the discovery of which was really made by Haller and others. In 1746, he published a duodecimo volume of letters on the mineral waters of Bearn.[17] De Bordeau imagined that he had discovered a duct connecting the thyroid gland and trachea. His name is attached to a theory, which states that each organ of the body produces and secrets a humor into the circulation.[18]

Camper

In 1760, Peter Camper, a distinguished Dutch physician and surgeon, published in Amsterdam his magnificent work on anatomy titled *Demonstrationum Anatomico-pathologicarum*,[19] the plates of which are original and remarkable for their accuracy. However, his death interrupted the completion of his work. A short time before his death, Camper published an account of his method of performing the operation of lithotomy in a Dutch journal based in Amsterdam. He said it had been communicated to him by M. Louis, who had informed him that he had not lost a patient by the operation since he had adopted it. However, there was little novelty in the method, which had been introduced by Peter Franco who had died nearly three centuries before. Around 1774, Camper communicated an account of the obstetric lever of Rouenhuysen[20] to the Royal Academy of Surgeons in Paris. As a result of this communication, the Academy admitted him as a foreign associate. Camper's

[15] Tubbs RS, et al. Emanuel Swedenborg (1688-1772), pioneer of neuroanatomy. Child's Nervous System 2011; 27(8), 1353-5.
[16] Recherches sur le pouls par rapport aux crises. 12mo. Paris, 1756.
[17] Lettres contenant des essais sur l'histoire des eaux minerals du Béarn, &c. 12mo. 1746.
[18] Bartolucci SL, Forbis P. Stedman's Medical Eponyms, 2nd edition. Baltimore, Lippincott Williams & Wilkins, 2005.
[19] Demonstrationum Anatomico-pathologicarum. Liber primus, in folio maximo cum quatuor figuris, and in 1762, Liber secundus, continens pelvis humanæ fabricam et morbos. Folio. Amstelodami.
[20] Rouenhuysen was an obstetrician in Amsterdam in the mid-to-late seventeenth century. Little information is available on his life and works. We know that Hugh Chamberlen sold his family secret of safe delivery to him for a significant amount of money. See, Lusk WT. The Science and Art of Midwifery. New York, Appleton, 1884, p 335.

name is still remembered by the eponym "Camper's fascia" given to the fatty outer layer of the superficial fascia of the anterior abdominal wall.[21]

A work in Latin on the fracture of the patella and olecranon by Camper was published posthumously by his son Adrian Gilles Camper in quarto at The Hague in 1789. This work was enriched with cases and illustrations from various authors and accompanied by two plates, in the last of which a representation is given of what is called a perfect cure of a fractured patella, although the two portions of the bone remained separated by a space at least four inches in length.

John Hunter

John Hunter (Figure 45), the younger brother of William Hunter, is regarded as the founder of scientific surgery.[22] He moved to London to assist his brother William in preparing and dissecting cadavers at his school of anatomy.[23] He also gave lectures and became as prominent of a public lecturer as his brother. Both brothers acquired great celebrity through establishing splendid museums.

Figure 45. John Hunter, 1728-1793 (Image courtesy of Reynolds Historical Library, the University of Alabama at Birmingham).

[21] Mian A, Bertino F, Shipley E, Shoja MM, Tubbs RS, Loukas M. Petrus Camper, A history and overview of the clinical importance of Camper's fascia in surgical anatomy. Clinical Anatomy 2013, doi, 10.1002/ca.22236.

[22] Maxwell B. John Hunter, the first surgical scientist. Can J Surg 1987; 30, 68-71.

[23] Evans CH. John Hunter and the origins of modern orthopaedic research. J Orthop Res 2007; 25, 556-60.

The Hunterian Museum of the Royal College of Surgeons in London harbors many of these specimens. John Hunter published the first and second part of his work on the natural history of the human teeth in London in 1771 and 1778, respectively.[24] His work on syphilis was published in 1786. John Hunter died suddenly from while speaking to one of his colleagues at St. George's Hospital in London on October 16, 1793, at the age of 66 years. On autopsy, his heart was found to be reduced in size, the coronary arteries completely calcified, and the valves undergoing calcification. He appears to have surpassed his brother in the impact of his investigations.

Cruikshank

William Cumberland Cruikshank, a celebrated anatomist,[25] was born at Edinburgh in 1746. For many years, he served William Hunter as a librarian, and later became his assistant lecturer. Cruikshank published *Anatomy of the Absorbent Vessels of the Human Body* in 1786.[26] In this work, he described the structure and distribution of the valvular lymphatic vessels. He substantially enriched Dr. Hunter's museum by contributing his beautiful preparations, especially by his curious injections of the lymphatics. In 1795, he communicated a valuable paper on the regeneration of the nerves to the Royal Society, which can be found in the *Philosophical Transactions* of that year. He also published a pamphlet on insensible perspiration in 1795 and an account of the appearances of the ovaries of rabbits at different stages of pregnancy in 1797. However, he will be remembered for his work on the lymphatics, which is indisputably the most accurate work on the subject for its time. Cruikshank died on June 27, 1800.

Baillie and William Hunter's Anatomy of the Gravid Uterus

The work *Anatomy of the Gravid Uterus*[27] was left incomplete because of the death of its author, William Hunter, in March 1783. It was published posthumously in London in 1775. This magnificent work describes all the principal changes that occur during the nine months of pregnancy in a series of 34 folio plates. It contains the first representation of the retroverted uterus and the membrana desidua reflexa, which was discovered by William Hunter. However, as William Hunter did not live to complete the anatomical descriptions of the figures, his nephew, Matthew Baillie, eventually completed them in 1794.[28,29]

---

[24] On the Natural History of the Teeth. 4to. London, 1771.

[25] Cruikshank studied at Edinburgh and Glasgow Universities, and at first he intended to become a priest. Aside from being an avid anatomist, Cruikshank was an able chemist. See, William Cumberland Cruikshank, a link with Samuel Johnson. Ann R Coll Surg Engl 1951; 8, 325-7.

[26] Anatomy of the absorbent vessels of the human body. London, 1786.

[27] The anatomy of the gravid uterus. London, 1775.

[28] Anatomical description of the gravid uterus and its contents. London, 1794.

[29] Dunn PM. Dr William Hunter (1718-83) and the gravid uterus. Arch Dis Child Fetal Neonatal Ed 1999; 80, F76-77.

Bell

John Bell was a distinguished surgeon and anatomist. He received his medical degree from the medical school in Edinburgh in 1779. He was a student of Alexander Monro secundus.[30] Bell published an excellent work on anatomy in four volumes.[31] This work exhibits the most complete and comprehensive overview of the progress and discoveries in anatomy up to the period of its publication. Only the first two of these four volumes were published in the eighteenth century, of which the first, published in 1793, discusses osteology and myology, and the second, published in 1797, discusses the anatomy of the heart and arteries. For his services to the field, Bell is rightfully called the "Father of Surgical Anatomy".

We have now reached the end of this book, which covered the history of medicine and surgery from the creation of the world to the commencement of the nineteenth century. We have endeavored to exhibit this history as fully and as faithfully as the scanty sources of information within our reach would permit us to do so.

[30] Kaufman MH. John Bell (1763-1820), the "father" of surgical anatomy. J Med Biogr 2005; 13, 73-81.

[31] Anatomy of the Human Body. vol. i. Anatomy of the bones, muscles, and joints. Edinburgh, 1793. vol. ii. Anatomy of the heart and arteries. Edinburgh, 1797. vol. iii. containing the nervous system. London, 1803. vol. iv. containing the abdominal viscera, &c. London, 1804, all in large octavo.

Author Index

A

B

V

W

Y

Z

Index

A

B

C

D

E

U

V

W